Advances in Intelligent Systems and Computing

Volume 220

Series Editor

J. Kacprzyk, Warsaw, Poland

For further volumes:
http://www.springer.com/series/11156

Jorge Casillas · Francisco J. Martínez-López
Rosa Vicari · Fernando De la Prieta
Editors

Management Intelligent Systems

Second International Symposium

 Springer

Editors

Jorge Casillas
Dept. Computer Science and
    Artificial Intelligence
University of Granada
Granada
Spain

Francisco J. Martínez-López
Dept. Business Administration
University of Granada
Granada
and
Marketing Group,
Open University of Catalonia
Barcelona
Spain

Rosa Vicari
Department of Computer Systems
University of Sao Paulo
Sao Paulo
Brazil

Fernando De la Prieta
Department of Computing Science
University of Salamanca
Salamanca
Spain

ISSN 2194-5357 ISSN 2194-5365 (electronic)
ISBN 978-3-319-00568-3 ISBN 978-3-319-00569-0 (eBook)
DOI 10.1007/978-3-319-00569-0
Springer Cham Heidelberg New York Dordrecht London

Library of Congress Control Number: 2013937324

Printed on acid-free paper

Springer is part of Springer Science+Business Media (www.springer.com)

Preface

This symposium was born as a research forum to present and discuss original, rigorous and significant contributions on Artificial Intelligence-based (AI) solutions—with a strong, practical logic and, preferably, with empirical applications—developed to aid the management of organizations in multiple areas, activities, processes and problem-solving; what we call Management Intelligent Systems (MiS).

Basically, the AI core focuses on the development of valuable, automated solutions (mainly by means of intelligent systems) to problems that would require the human application of intelligence. In an organizational context, there are problems that require human judgment and analysis to assess and solve these problems with a guarantee of success. These decisional situations frequently relate to strategic issues in organizations in general, and in firms in particular, where problems are far from being well structured. In essence, AI offers real opportunities for advancing the analytical methods and systems used by organizations to aid their internal and external administration processes and decision-making. Indeed, well-conceived and designed intelligent systems are expected to outperform operational research- or statistical-based supporting tools in complex, qualitative and/or difficult-to-program administration problems and decisional scenarios. However, these opportunities still need to be fully realized by researchers and practitioners. Therefore, more interdisciplinary and applied contributions are necessary for this research stream to really take off.

Each paper submitted to IS-MiS 2013 went through a stringent peer review process by members of the Program Committee comprising 54 internationally renowned researchers (including a dozen Editors-in-Chief of prestigious research journals on management, business and intelligent systems) from 16 countries. The quality of papers was on average good, with an acceptance rate of approximately 70%.

This volume also includes a track focused on the latest research on Intelligent Systems and Technology Enhanced Learning (iTEL), as well as its impacts for learners and institutions. It aims at bringing together researchers and developers from both the professional and the academic realms to present, discuss and debate the latest advances on intelligent systems and technology-enhanced learning. The program committee of iTEL is composed of 12 members from 7 countries; all the contributions have passed a blind-review process with an average acceptance rate around 75%.

We would like to thank all the contributing authors, the reviewers, the sponsors (IEEE Systems Man and Cybernetics Society Spain, AEPIA Asociación Española para la Inteligencia Artificial, APPIA Associação Portuguesa Para a Inteligência Artificial, CNRS Centre national de la recherché scientifique), as well as the members of both the Program Committee and the Organising Committee. All these people's efforts have contributed to the success of the IS-MiS and iTEL.

The Editors
Jorge Casillas
Francisco J. Martínez-López
Rosa Vicari
Fernando De la Prieta

Organization

General Chair

Jorge Casillas   University of Granada, Spain

Program Chair

Francisco J. Martínez López   University of Granada and Open University
                                of Catalonia, Spain

Program Committee

Andrea Ahlemeyer-Stubbe   antz21, Germany
Daniel Arias   University of Granada, Spain
Barry J. Babin   Louisiana Tech University, USA
P.V. Sundar Balakrishnan   University of Washington, USA
Malcolm J. Beynon   Cardiff University, UK
Min-Yuang Cheng   National Taiwan University of Science and
                                Technology, Taiwan

Bernard De Baets   University of Ghent, Belgium
Anastasios Doulamis   Technical University of Crete, Greece
Flavius Frasincar   Erasmus University Rotterdam, The Netherlands
Bob Galliers   Bentley University, USA
Juan Carlos Gázquez-Abad   University of Almería, Spain
Dawn Iacobucci   Vanderbilt University, USA
Frank Klawonn   Ostfalia University of Applied Sciences, Germany
Kemal Kiliç   Sabanci University, Turkey
Khairy A. H. Kobbacy   University of Salford, UK
Subodha Kumar   Texas A&M University, USA
Dalia Kriksciuniene   Vilnius University, Lithuania
Peter LaPlaca   University of Hartford, USA
Nick Lee   Aston Business School, UK

Track on Intelligent Technologies for Enhanced Learning

Ana Belén Gil González	University of Salamanca, Spain
Silvana Aciar	Instituto de Informática, Universidad Nacional de San Juan, Argentina
Jon Mikel	Lund University, Sweden
Néstor Darío Duque M	Universidad Nacional de Colombia, Colombia
Eliseo Reategui	UFRGS, Brazil
Amparo Jiménez	Pontifical University of Salamanca(Spain
Yaxin Bi	University of Ulster, Ireland

Local Organising Committee

Juan M. Corchado	University of Salamanca, Spain
Javier Bajo	Pontifical University of Salamanca, Spain
Juan F. De Paz	University of Salamanca, Spain
Sara Rodríguez	University of Salamanca, Spain
Dante I. Tapia	University of Salamanca, Spain
Fernando De la Prieta	University of Salamanca, Spain
Davinia Carolina Zato Domínguez	University of Salamanca, Spain
Gabriel Villarrubia González	University of Salamanca, Spain
Alejandro Sánchez Yuste	University of Salamanca, Spain
Antonio Juan Sánchez Martín	University of Salamanca, Spain
Cristian I. Pinzón	University of Salamanca, Spain
Rosa Cano	University of Salamanca, Spain
Emilio S. Corchado	University of Salamanca, Spain
Eugenio Aguirre	University of Granada, Spain
Manuel P. Rubio	University of Salamanca, Spain
Belén Pérez Lancho	University of Salamanca, Spain
Angélica González Arrieta	University of Salamanca, Spain
Vivian F. López	University of Salamanca, Spain
Ana de Luís	University of Salamanca, Spain
Ana B. Gil	University of Salamanca, Spain
Mª Dolores Muñoz Vicente	University of Salamanca, Spain
Jesús García Herrero	University Carlos III of Madrid, Spain

Contents

2nd International Symposium on Management Intelligent Systems (ISMiS'13)

1st International Workshop on Intelligent Technologies for Enhanced Learning (iTEL'13)

Making Accurate Credit Risk Predictions with Cost-Sensitive MLP Neural Networks

R. Alejo, V. García, A.I. Marqués, J.S. Sánchez, and J.A. Antonio-Velázquez

Abstract. In practical applications to credit risk evaluation, most prediction models often make inaccurate decisions because of the lack of sufficient default data. The challenging issue of highly skewed class distribution between defaulter and non-defaulters is here faced by means of an algorithmic solution based on cost-sensitive learning. The present study is conducted on the popular Multilayer Perceptron neural network using three misclassification cost functions, which are incorporated into the training process. The experimental results on real-life credit data sets show that the proposed cost functions to train such a neural network are quite effective to improve the prediction of examples belonging to the defaulter (minority) class.

1 Introduction

Credit scoring and behavioral management tools have been used by financial institutions to distinguish the creditworthy credit applicants from those who will probably default with their repayments [17]. The most classical approaches to credit scoring rely on statistical and operations research methods, but one can also find more advanced techniques that belong to the field of soft computing, such as neural net-

R. Alejo · J.A. Antonio-Velázquez
Tecnológico de Estudios Superiores de Jocotitlán, Carretera Toluca-Atlacomulco,
Col. Ejido de San Juan y San Agustn, 50700 Jocotitlán, Mexico
e-mail: ralejoll@hotmail.com

V. García · J.S. Sánchez
Institute of New Imaging Technologies, Department of Computer Languages and Systems,
Universitat Jaume I, Av. Sos Baynat s/n, 12071 Castelló de la Plana, Spain
e-mail: {jimenezv, sanchez}@uji.es

A.I. Marqués
Department of Business Administration and Marketing, Universitat Jaume I, Av. Sos Baynat
s/n, 12071 Castelló de la Plana, Spain
e-mail: imarques@uji.es

J. Casillas et al. (Eds.): *Management Intelligent Systems*, AISC 220, pp. 1–8.
DOI: 10.1007/978-3-319-00569-0_1   © Springer International Publishing Switzerland 2013

works, support vector machines, fuzzy systems, and evolutionary algorithms. The credit risk prediction models are built on historical data (collected by banks and/or financial institutions) using samples that describe socio-demographic characteristics and economic conditions of previous credit applicants, and the outputs are scores that help lenders to classify a new applicant as "good" or "bad" depending on the probability of defaulting on repayments.

According to the Basel II Committee, the historical data may not be reliable enough to produce accurate scores when there is not available a sufficient number of examples of the defaulter class. In practice, however, there are a wide variety of business applications where the number of defaulters is less than 10% of the whole data [8]. This situation is usually referred to as the low-default portfolio problem or the class imbalance problem [10]. A two-class data set is said to be imbalanced when one of the classes (the minority one) is heavily under-represented as regards the other (the majority) class. It has been observed that class imbalance may have much influence on the performance of conventional credit risk prediction models because they implicitly assume that the class prior probabilities and the misclassification costs are equal [2, 8]. This issue is particularly important for financial applications such as bankruptcy prediction [1], credit card fraud detection [18] and creditworthiness evaluation [12] since it may be very costly to misclassify minority samples.

In credit scoring, learning from imbalanced data sets has usually been tackled by resampling the data, either by over-sampling the minority class and/or under-sampling the majority class until both classes are similar in size [2, 3, 5, 11, 16, 20]. Another typical approach refers to cost-sensitive learning [9, 14, 15, 19], which consists of the inclusion of costs into the learning algorithm for strengthening the discrimination process towards the minority class.

This paper explores three different methods to define misclassification costs that will be further incorporated into Multilayer Perceptron (MLP) neural networks for credit risk evaluation problems. Significant results obtained from a thorough experimental study on 25 imbalanced credit data sets prove the performance gains when introducing misclassification costs into the learning algorithm, thus leading to important advantages to detect fraudulent financial activities and accurately predict creditworthiness of applicants. This may provide decision makers with efficient and effective tools to assess credit risk more precisely.

2   Cost-Sensitive MLP

The most popular learning procedure for the MLP neural network is the backpropagation algorithm, which uses a set of training instances for adjusting the free parameters U. Several works have shown that the class imbalance problem generates unequal contributions to the mean square error (MSE) during the training phase, where the major contribution to the MSE is produced by the majority class.

Given a training data set with two classes ($J = 2$) of size $N = \sum_j^J n_j$, where n_j is the number of samples in class j, the MSE for class j can be expressed as

$$E_j(U) = \frac{1}{N} \sum_{n=1}^{n_j} (t^n - z^n)^2,$$ (1)

where t^n is the desired output and z^n is the actual output of the network for the sample n.

Then the overall MSE can be written in terms of $E_j(U)$ as follows:

$$E(U) = \sum_{j=1}^{J} E_j(U) = E_1(U) + E_2(U).$$ (2)

When $n_1 << n_2$, then the $E_1(U) << E_2(U)$ and $\|\nabla E_1(U)\| << \|\nabla E_2(U)\|$, where the operator ∇ denotes the gradient of the error function. Consequently, $\nabla E(U) \approx \nabla E_2(U)$. So, $-\nabla E(U)$ is not always the best direction to minimize the MSE in both classes.

The unequal contribution to the MSE can be compensated by introducing a cost function (γ) in order to avoid the MLP ignores the minority class:

$$E(U) = \sum_{j=1}^{J} \gamma(j)E_j = \gamma(1)E_1(U) + \gamma(2)E_2(U)$$

$$= \frac{1}{N} \sum_{j=1}^{J} \gamma(j) \sum_{i=1}^{n_j} \sum_{p=1}^{J} (t_p^i - z_p^i)^2,$$ (3)

where $\gamma(1)\|\nabla E_1(U)\| \approx \gamma(2)\|\nabla E_2(U)\|$.

In this work, we define three different cost functions: (i) CS-MLP1, $\gamma(j) = n_{max}/n_j$, where n_{max} is the number of majority class samples; (ii) CS-MLP2, $\gamma(j) = N/n_j$; and (iii) CS-MLP3, $\gamma(j) = \|\nabla E_{max}(U)\|/\|\nabla E_j(U)\|$, where $\|\nabla E_{max}(U)\|$ corresponds to the majority class.

3   Experimental Protocol

Following the experimental set-up used in other papers [10, 11], five real-life credit data sets have been taken to test the performance of the strategies investigated in the present paper. The widely-used Australian, German and Japanese data sets are from the UCI Machine Learning Database Repository [6]. The UCSD data set corresponds to a reduced version of a database used in the 2007 Data Mining Contest organized by the University of California San Diego and Fair Isaac Corporation. The Iranian data set [13] comes from a modification to a corporate client database of a small private bank in Iran.

Each original set, except the Iranian database because of its extremely high imbalance ratio (*iRatio* = 19), has been altered by randomly under-sampling the minority class of defaulters, thus producing six data sets with varying imbalance ratios, *iRatio* = {4, 6, 8, 10, 12, 14}. Therefore, we have obtained a total of 25

Table 1 Some characteristics of the data sets used in the experiments. Note that all input variables were represented as numeric values

Data set	#Attributes	#Examples	#Good	#Bad	iRatio
Australian4	14	384	307	77	4
German4	24	875	700	175	4
Japanese4	15	370	296	74	4
UCSD4	38	2295	1836	459	4
Australian6	14	358	307	51	6
German6	24	817	700	117	6
Japanese6	15	345	296	49	6
UCSD6	38	2142	1836	306	6
Australian8	14	345	307	38	8
German8	24	788	700	88	8
Japanese8	15	333	296	37	8
UCSD8	38	2066	1836	230	8
Australian10	14	338	307	31	10
German10	24	770	700	70	10
Japanese10	15	326	296	30	10
UCSD10	38	2020	1836	184	10
Australian12	14	333	307	26	12
German12	24	758	700	58	12
Japanese12	15	321	296	25	12
UCSD12	38	1989	1836	153	12
Australian14	14	329	307	22	12
German14	24	750	700	50	12
Japanese14	15	317	296	21	12
UCSD14	38	1967	1836	131	12
Iranian19	27	1000	950	50	19

data sets. Table 1 summarizes the main characteristics of the data sets, including the imbalance ratio, that is, the number of non-default examples divided by the number of default cases.

We have adopted a 5-fold cross-validation method to estimate the performance: each data set has been split into five stratified blocks or folds of size $N/5$ (where N denotes the total number of examples in the data set). Subsequently, five iterations of training and test are performed such that within each iteration a different fold of the data is held-out for test while the remaining four folds are used for training the classifier. Accordingly, the results correspond to the average over the five runs. The four different MLP models have been applied to the imbalanced training sets. For each MLP consisting of one hidden layer with four neurons, the learning rate and the momentum have been set at 0.1 and 0.01 respectively, whereas the stopping criterion has been fixed to either 25,000 epochs or MSE $= 0.001$.

3.1 Performance Evaluation Metric

Most credit scoring applications often employ classification accuracy (Acc) and/or error rates to estimate the performance of learning systems. However, empirical and theoretical evidences show that these measures are biased with respect to data imbalance and proportions of correct and incorrect classifications. To face with the class imbalance problem, the receiver operating characteristic (ROC) curve has been suggested as a suitable tool for visualizing and selecting classifiers based on their trade-offs between benefits (true positives) and costs (false positives). A quantitative representation of a ROC curve is the area under it (AUC). For just one run of a classifier, the AUC can be computed as $AUC = (sensitivity + specificity)/2$, where *sensitivity* is the percentage of bad examples that have been predicted correctly, whereas *specificity* corresponds to the percentage of good instances predicted as good.

3.2 Statistical Tests

The AUC results have further been tested for statistically significant differences among the MLP models by means of the Iman-Davenport's statistic [4, 7] at significance levels of 5% and 10%. The process starts by computing the Friedman's ranking of the algorithms for each data set independently according to the AUC results: as there are four competing strategies, the ranks for each data set go from 1 (best) to 4 (worst); in case of ties, mean ranks are assigned. Then the average rank of each algorithm across all data sets is computed. Under the null-hypothesis, which states that all the algorithms are equivalent, the Friedman's statistic can be computed as follows:

$$\chi_F^2 = \frac{12N}{K(K+1)} \left[\sum_j R_j^2 - \frac{K(K+1)^2}{4} \right] \qquad (4)$$

where N denotes the number of data sets, K is the total number of algorithms, and R_j is the average rank of the algorithm j.

As the Friedman's test produces an undesirably conservative effect [7], the Iman-Davenport's statistic constitutes a better alternative. This is distributed according to the F-distribution with $K-1$ and $(K-1)(N-1)$ degrees of freedom:

$$F_F = \frac{(N-1)\chi_F^2}{N(K-1) - \chi_F^2} \qquad (5)$$

If the null-hypothesis of equivalence is rejected, we can then proceed with a post hoc test. In this work, the Holm's post hoc test has been employed to ascertain whether the best (control) algorithm performs significantly better than the remaining techniques [7].

4 Experimental Results and Discussion

Table 2 shows the AUC values across all data sets when using the four different MLP models. In general, the three cost-sensitive MLP classifiers usually perform better than the original MLP (without a cost function γ), what can be seen by either analyzing the global AUC or comparing the AUC of each algorithm over each data set. The Friedman's average ranks for the four classification models have been plotted in Fig. 1 (a), supporting the previous assertion of the findings with the AUC values where the use of the traditional MLP produces the highest (worst) average

Table 2 Average AUC values (the best result for each database is highlighted in bold face)

Data sets	MLP	CS-MLP1	CS-MLP2	CS-MLP3
Australian4	0.7783	0.7965	**0.8009**	0.7985
German4	0.6392	0.6639	**0.6667**	0.6587
Japanese4	**0.8332**	0.8126	0.8151	0.8311
UCSD4	0.7622	**0.8040**	0.8006	0.8035
Australian6	0.7173	**0.7396**	0.7351	0.7268
German6	0.6202	**0.6503**	0.6486	0.6464
Japanese6	0.7623	0.7472	0.7346	**0.7759**
UCSD6	0.6850	0.7898	**0.7943**	0.7708
Australian8	**0.7469**	0.6965	0.6998	0.7399
German8	0.5900	**0.6430**	0.6314	0.6355
Japanese8	0.7650	0.8184	**0.8337**	0.8257
UCSD8	0.6584	**0.7866**	0.7845	0.7752
Australian10	0.6159	0.6218	0.6156	**0.6298**
German10	0.5776	0.5985	0.5805	**0.6094**
Japanese10	0.6259	0.6446	**0.6747**	0.6252
UCSD10	0.5976	**0.7860**	0.7845	0.7649
Australian12	0.6558	0.7131	0.6836	**0.7346**
German12	0.5380	0.5697	**0.5796**	0.5501
Japanese12	0.6811	**0.7453**	0.7301	0.7245
UCSD12	0.6072	0.7670	**0.7697**	0.7684
Australian14	0.6180	0.6558	0.6758	**0.6798**
German14	0.5669	**0.5904**	0.5770	0.5813
Japanese14	0.6333	0.6166	0.6273	**0.6347**
UCSD14	0.5853	0.7559	**0.7569**	0.7432
Iranian19	0.5001	**0.6439**	0.6224	0.6038
Average	0.6544	**0.7063**	0.7049	0.7055

ranks. Also, it is worth noting that the CS-MLP1 strategy clearly arises as the best approach with the lowest ranking, that is, the highest AUC value.

With the aim of checking whether our conclusions can be supported by non-parametric statistical tests, the Iman-Davenport's statistic has been computed to discover whether or not the AUC results are significantly different. This computation yielded $F_F = 9.3333$ distributed according to F-distribution with 3 and 72 degrees of freedom. The p-value obtained using $F(3, 72)$ is $2.7354029E - 5$, so the null-hypothesis that all strategies here explored perform equally well can be rejected at both significance levels of $\alpha = 0.10$ and $\alpha = 0.05$. Consequently, we can now carry on with a Holm's post hoc test, using the CS-MLP1 classifier as the control algorithm.

The results of the Holm's post hoc test have been plotted in Fig 1(b), whereby all strategies have been sorted according to their p-values. Both the dot and the single lines represent the α, which was adjusted by dividing it by the number of comparisons made (α/(k-i)) when $\alpha = 0.10$ and $\alpha = 0.05$. The Holm's procedure rejects the first hypothesis since the corresponding p-value is smaller than the adjusted α. This suggests that CS-MLP1 is only superior to the MLP and there are not significance differences with the remaining cost-sensitive strategies. An additional Friedman's test carried out among the three cost-sensitive MLP models confirms the previous statements, where the computed p-value was 0.763137835.

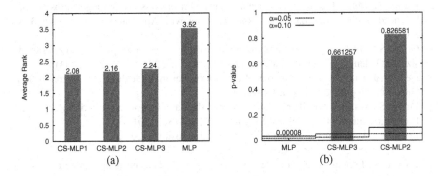

Fig. 1 Friedman's rankings (a) and Holm's post hoc test (b)

5 Conclusions

This paper has focused on the cost-sensitive learning approach to credit risk prediction with asymmetric misclassification costs and skewed class proportion between defaulters and non-defaulters. Three cost functions have been introduced into the learning process of a MLP neural network trained with the back-propagation algorithm. The aim of these cost functions is to compensate the unequal contribution of the MSE during the training phase.

Experimental results over 25 real-life imbalanced credit data sets have demonstrated that the cost-sensitive models achieve significant improvements in terms of the AUC measure when compared to the original MLP (without a cost function), especially with the CS-MLP1 approach because its cost function appears to better correct the biased behavior towards the majority class in the MSE. Taking this into account, the use of cost-sensitive MLP may be of great relevance for banks and financial institutions since a small increase in performance may result in significant future savings and have important managerial/business implications.

Future work will be mainly addressed to explore the application of cost-sensitive learning to other types of neural networks in order to handle the class imbalance problem present in most credit risk prediction problems.

Acknowledgements. This work has partially been supported by the Spanish Ministry of Education and Science and the Generalitat Valenciana under grants TIN2009–14205 and PROMETEO/2010/028, respectively.

References

1. Alfaro-Cid, E., Sharman, K.C., Esparcia-Alcázar, A.I.: A genetic programming approach for bankruptcy prediction using a highly unbalanced database. In: Giacobini, M. (ed.) EvoWorkshops 2007. LNCS, vol. 4448, pp. 169–178. Springer, Heidelberg (2007)

2. Brown, I., Mues, C.: An experimental comparison of classification algorithms for imbalanced credit scoring data sets. Expert Systems with Applications 39(3), 3446–3453 (2012)

3. Crone, S.F., Finlay, S.: Instance sampling in credit scoring: An empirical study of sample size and balancing. International Journal of Forecasting 28(1), 224–238 (2003)

4. Demšar, J.: Statistical comparisons of classifiers over multiple data sets. Journal of Machine Learning Research 7(1), 1–30 (2006)

5. Finlay, S.: Multiple classifier architectures and their application to credit risk assessment. European Journal of Operational Research 210, 368–378 (2011)

6. Frank, A., Asuncion, A.: UCI machine learning repository (2010),
   `http://archive.ics.uci.edu/ml`

7. García, S., Fernández, A., Luengo, J., Herrera, F.: Advanced nonparametric tests for multiple comparisons in the design of experiments in computational intelligence and data mining: Experimental analysis of power. Information Sciences 180(10), 2044–2064 (2010)

8. Hand, D.J., Vinciotti, V.: Choosing k for two-class nearest neighbour classifiers with unbalanced classes. Pattern Recognition Letters 24(9-10), 1555–1562 (2003)

9. Harris, T., Gittens, C.: Minimising expected misclassification cost when using support vector machines for credit scoring. In: Proc. 7th International Multi-Conference on Computing in the Global Information Technology, pp. 225–231 (2012)

10. Kennedy, K., Mac Namee, B., Delany, S.J.: Learning without default: A study of one-class classification and the low-default portfolio problem. In: Coyle, L., Freyne, J. (eds.) AICS 2009. LNCS, vol. 6206, pp. 174–187. Springer, Heidelberg (2010)

11. Marqués, A.I., García, V., Sánchez, J.S.: On the suitability of resampling techniques for the class imbalance problem in credit scoring. Journal of the Operational Research Society (2012), doi:10.1057/jors.2012.120

12. Paleologo, G., Elisseeff, A., Antonini, G.: Subagging for credit scoring models. European Journal of Operational Research 201, 490–499 (2010)

13. Sabzevari, H., Soleymani, M., Noorbakhsh, E.: A comparison between statistical and data mining methods for credit scoring in case of limited available data. In: Proc. the 3rd CRC Credit Scoring Conference (2007)

14. Schebesch, K.B., Stecking, R.: Support vector machines for credit scoring: extension to non standard cases. In: Proc. of the 27th Annual Conference of the Gesellschaft fur Klassikation e. V., pp. 443–451 (2003)

15. Stecking, R., Schebesch, K.B.: Classification of large imbalanced credit client data with cluster based SVM. In: Proc. of the 34th Annual Conference of the Gesellschaft fur Klassikation e. V., pp. 443–451 (2008)

16. Thai-Nghe, N., Nghi, D.T., Schmidt-Thieme, L.: Learning optimal threshold on resampling data to deal with class imbalance. In: Proc. IEEE RIVF International Conference on Computing and Telecommunication Technologies, pp. 71–76 (2010)

17. Thomas, L.C., Edelman, D.B., Crook, J.N.: Credit Scoring and Its Applications. SIAM, Philadelphia (2002)

18. Vinciotti, V., Hand, D.J.: Scorecard construction with unbalanced class sizes. Journal of the Iranian Statistical Society 2(2), 189–205 (2003)

19. Yao, P.: Comparative study on class imbalance learning for credit scoring. In: Proc. 9th International Conference on Hybrid Intelligent Systems, vol. 2, pp. 105–107 (2009)

20. Zhang, L., Wang, W.: A re-sampling method for class imbalance learning with credit data. In: Proc. the 2011 International Conference on Information Technology, Computer Engineering and Management Sciences, pp. 393–397 (2011)

Multilevel Clustering on Very Large Scale of Web Data

Amine Chemchem and Habiba Drias

Abstract. With the evolution of the WWW, the computer world has become a huge wave of data, to perform a search of this data, the classical approaches of data mining are still valid, but with diminished performance. In this paper, we present a new clustering approach based on multilevel paradigm called multilevel clustering, that allows to divert the complexity of calculation and execution period of data mining on very large scale. The developed algorithm have been implemented on three public benchmarks to test the effectiveness of the multilevel clustering approach. The numerical results have been compared to those of the simple k-means algorithm. As foreseeable, the multilevel clustering outperforms clearly the basic k-means on both the execution time and success rate that reaches 100 % while increasing the number of data.

1 Introduction

Data clustering is well-known as an important tool in data analysis. It uses data similarity measures to partition a large data set into a set of disjoint data clusters, such as data points within the clusters are close to each other, and the data points from different clusters are dissimilar from each other [1]. Also it is widely recognized that numerical data clustering differs from categorical data clustering in terms of the types of data similarity measures used. Nowadays, in Web, data are generally not homogenous. So, it can includes all types of data mentioned previously.

One possible idea for processing this data; is to transform the web data set to a traditional categorical, numeric, or boolean data set, but the large volume and high dimensionality of the transformed data set make the existing algorithms inefficient to process this last. For this reason our idea is to process directly the web data set without any transformation or pretreatment step. The most difficult problem for

Amine Chemchem · Habiba Drias
USTHB-LRIA, BP 32 El Alia Bab Ezzouar, Algiers, Algeria
e-mail: aminechemchem@gmail.com, hdrias@usthb.dz

J. Casillas et al. (Eds.): *Management Intelligent Systems*, AISC 220, pp. 9–16.
DOI: 10.1007/978-3-319-00569-0_2   © Springer International Publishing Switzerland 2013

clustering this data is to find the appropriate distance measure, and also to define a formula for calculate the gravity centers of the resultants clusters.

The remainder of this document is organized as follows. The next section presents the concept of clustering data, with the proposed formulas of similarity measure, and centroids computation. Afterwards, the multilevel paradigm is explained in section 3. Then a new clustering approach based on multilevel paradigm for web data is presented in section 4. The experimental evaluation presented in section 5 is performed on a benchmark including three public data bases. Conclusions are finally summarized and some perspectives are suggested.

2   Simple Clustering of Web Data

In the history of clustering in data mining technics, there is an increasing interest in the use of clustering methods in different domains[1], for example we can cite clustering in pattern recognition [5], image processing [6] and information retrieval [7]. Clustering has a rich history in other disciplines [8] such as biology, psychiatry, psychology, archaeology, geology, geography, and marketing. In this work, we are interesting by data clustering, and here also it exists different kinds of data; we can cite the work of P.Aggrawal et al [10] that presents a new similarity measures between two categorical attributes in a relational database.

In our work we try to ignore the kind of processed data, and start the clustering without any step for data pretreatment, one of applications of our present work can be used to define the domain of processed data.

2.1   Distance Measure

Many distance measures have been proposed in literature of data mining in order to classify or to cluster homogeneous data. Most often, these measures are metric functions; Manhattan distance, Minkowski distance and Hamming distance. Jaccard index, Cosine Similarity and Dice Coefficient are also popular distance measures. For non-numeric data sets, special distance functions are proposed. For example, edit distance is a well-known distance measure for text attributes[9]. But nowadays the web is characterized by huge data bases from different fields, and then these classical similarity measures are inefficient.

In this work we are interested by giving a similarity measure function that can be applied on all kind of data (numerical, categorical, transactional,...), and here also, we can cite a few interesting works [2][4][3], that propose a similarity measures formulas based on weighting the attributes of data, so we need to have a predeterminate knowledge on data treated before calculating the similarity between them, in order to obtain the correct distance [3], or to favoring one type of attribute to an other one like in [4].

Our idea is inspired by the way of human thinking; so if we have to know the difference between two things, we are not interested by the features of a single object, but we need to find just their shared and different features.

$$Sim\_C(data_i, data_j) = (\bigcup Clauses(data_i, data_j)) - (\bigcap Clauses(data_i, data_j))$$
$$\text{(1)}$$

Where: clauses($data_i$)= ($attribute_1, value_1$),($attribute_2, value_2$),...,($attribute_n, value_n$)
For example: if we have 2 data as follows:

$data_1$: age=28 years, $income = 3000\$$, maried=yes.
$data_2$: name=casillas, age=28 years, team=real madrid.
Clauses ($data_1$)=(age,28),(income,3000),(maried,yes).
Clauses ($data_2$)=(name,casillas),(age,28),(team,real madrid).
$Sim\_c(data_1, data_2)$= total clauses - shared clauses = $5 - 1 = 4$.

2.2 Gravity Center Computation

Our idea for gravity center calculation of each cluster is based on the frequency of
the features, so the length of the centroid must be equal to the average of lengths
of all data that belong to its cluster. And with the table of frequency, the centroid is
constructed with the most frequent features, as explained in the following example:
    If we have 3 data in the same cluster as follows:

$data_1$: weather=sunny, $temperature = 27$, $play\_tennis = yes$.
$data_2$: age=28 years, $income = 3000\$$, situation=married, children =1.
$data_3$: name=Casillas, age=28 years, situation=married, team=real Madrid.

From previous data we remark that their lengths are respectively 3,4,4, then the
centroid must be constructed from 4 different features (the average of lengths). For
that we choose the most frequent features; so the cendroid C is featured by:
    C ={(age, 28), (situation, married), (children, 1), (name, casillas)}.

3   Multilevel Clustering Web Data

The multilevel paradigm is a simple technique which at its core involves recursive
coarsening to produce smaller and smaller problems that are easier to solve than the
original one. It consists of three phases: coarsening, initial solution, and uncoars-
ening. The coarsening phase aims at merging the components associated with the
problem to form clusters. The clusters are used in a recursive manner to construct a
hierarchy of problems each representing the original problem but with fewer degrees
of freedom. The coarsest level can then be used to compute an initial solution. The
solution found at the coarsest level is extended to give an initial solution for the next
level, and then improved using a chosen optimization algorithm. A common feature
that characterizes multilevel algorithms is that any solution in any of the coarsened
problems is a legitimate solution to the original one.
    In this section, we develop a new general algorithm for web data clustering based
on multilevel paradigm. First, it begins by eliminating the nearest data from the
original data set in the coarsening step to keep only the most far away from each

other. Thanks to a genetic algorithm that computes at the end of this stage a mini data base that includes the different most distant data. After this step, the k-means algorithm is launched on the smaller data base and it is expectable that it will provide reliable centers of gravity, because the selected data are very distant from each other. Finally in the refinement step, the k-NN algorithm is invoked on the result of k-means of the previous step. This operation is repeated until obtaining the initial data set. The purpose of using k-NN is to complete inserting the remaining data in the corresponding classes.

3.1 Coarsening Step

The purpose of this step is to reduce the data number. We use a genetic algorithm for this aim while ensuring a good sweeping of all the data base. The data are not removed randomly but in a way to keep only those that appear very far from each other.

- A chromosome (individual): is an encoded solution, in our case it is represented by an array of integers varying between 0 and the size of data set, each gene representing one data. Therefore the length of the chromosome represents the chosen size of the smallest data base.
- Fitness function The notion of fitness is fundamental to the application of genetic algorithms. It is a numerical value that expresses the performance of an individual (solution), so that different individuals can be compared. We propose as a fitness function the sum of the distances between each of its genes, that is:
  fitness(chromosome) = MAX $(\sum_{i=1}^{n-1} \sum_{j=i+1}^{n}$ Sim_C(i,j)), for all i different from j, with i and j belonging to chromosome.

With the previous operators , the coarsening algorithm is presented (Algorithm1):

Algorithm 1. Genetic Algorithm for Coarsening Step

1: Generate randomly an initial population of solutions.
2: Calculate fitness(s) for each solution of the initial population
3: **while** i < Max-Iter **do**
4:    Generate *RC* and *RM*: a random integer between[1 and 100]/*crossover and mutation rate*/.
5:    **if** RC < rate of crossover **then**
6:       Apply the crossover (on 2 individuals chosen randomly).
7:    **end if**
8:    **if** RM < rate of mutation **then**
9:       Apply the mutation (on 1 individual chosen randomly).
10:   **end if**
11:   Evaluate fitness(S') /* where S' is the new individual*/
12:   Consider the best found solution.
13: **end while**

3.2 Initial Solution Step

After the coarsening step, we obtain a best solution that maximizes the distances between data. So it is sure that the last one sweep all the areas of the original data set. In this step, we apply a simple k-means algorithm to cluster these data and the reliable gravity centers will be achieved.

3.3 Refinement Step

After the clustering in the initial solution step, the different clusters are obtained with the data belonging to the initial solution. In order to rebuild the initial data set, we apply the k-nearest-neighbors algorithm for the rest of data, as shown in the following algorithm:

Algorithm 2. Refinement step algorithm based on K-NN

$data - list := alldata(database) - data(initialsolution)$.
while $data - list$ is not empty **do**
  Current_data := get_data (data-list).
  for each already classified data D_i **do**
    Calculate the distance Sim_C(current-data,D_i).
  end for
  Compute k_nearest_neighbor($current - data$)
  for each data belonging to k-nearest-neighbors **do**
    Calculate the number of frequency of each class.
  end for
  Attribute to "current-data" the most frequent class.
end while

---

4 Evaluation and Experimentation

In this part, we build our data set from three different benchmarks. It includes in all 45000 data:

- The first one is originally a big data set known as: **Chess (King-Rook vs. King) Data Set**[1]. It contains 28056 instances with 7.
- The second one is known as: **Abalone Data Set**[2], it contains 4177 instances with 9 attributes.
- The third data set is known as: **Letter Recognition Data Set**[3], it contains 20000 instances, with 17 attributes.

---

[1] http://archive.ics.uci.edu/ml/datasets/Chess+%28King-Rook+vs.+King%29

[2] http://archive.ics.uci.edu/ml/datasets/Abalone

[3] http://archive.ics.uci.edu/ml/datasets/Letter+Recognition

4.1   Evaluation Pattern

Our benchmark as it is presented in the previous subsection(A), is built from three different benchmarks. To evaluate the clustering success rate, we fixed the parameter k of the clustering algorithms to 3, then after the clustering step, the three obtained clusters are compared to the initial data bases on the basis of the number of maximum common data; that is means the clusters must contain all data that provide from the same data set, and than success rate is calculated using the following formula:

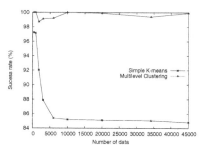

Fig. 1 Success Rate of Clustering Approaches

Success $rate = \frac{ncd}{npd} = \frac{ncd1}{npd1} + \frac{ncd2}{npd2} +$

$\frac{ncd3}{npd3}...$ (2) Where: $ncd = number$ of correct data. npd = number of pertinent data. The number of pertinent data of a $DB_i = The$ total number of its data.

4.2   Experimentation

4.2.1   Performance Comparison

Figure1 compares the performance of the designed approach based on multilevel paradigm to the simple k-means approach. We remark that the results computed by the simple k-means and the multi-level k-means are very different. While the success rate of the simple k-means algorithm is reduced from 97,5% to 85%, with increasing the number of data to cluster from 1000 to 45000data, the multilevel algorithm's success rate remain constant to 100%.

4.2.2   CPU Runtime Comparison

The execution time for the different clustering approaches are compared in Figure 2 and Figure 3.

We note in Fig.2, when increasing the number of web data, the multilevel clustering algorithm is faster than the simple K-means. Moreover, when the simple

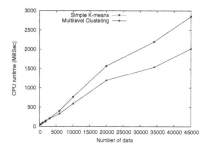

Fig. 2 CPU time vs rules number

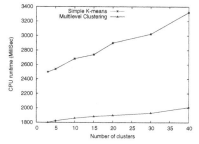

Fig. 3 CPU Time vs cluster number

clustering approach deals a number of data varying between 400 and 45000, the execution time increases from 55 to 2900 milliseconds whereas the execution time of multilevel approach increases just between 55 and 2000 milliseconds.

Fig.3 shows how the execution time change, with the increase of parameter "K" that represents the number of clusters, for the two approaches: simple clustering and the clustering based multilevel paradigm. According to this figure, Simple_K-means is slower than the multilevel clustering approach.Execution time of multilevel algorithm doesn't exceed 2000 ms, when the number of clusters is 40, even though simple k-means execution time is 3300 ms when the number of clusters is 40.

4.2.3 General Comparison

When comparing the two clustering approaches, we note that the approach based on multilevel paradigm is more efficient on both solution quality and runtime criteria than the simple clustering approach, so it takes more than 1300 milliseconds less time than the simple clustering approaches to deal the same data base, and we explained that by the fact that simple k-means repeat many iteration of computation the new gravity centers. On the contrary, the multilevel algorithm applies the K-means algorithm only on a much smaller data set, at its coarsest level. Furthermore, even on the quality criterion expressed through the success rate, its performance reaches 100% whatever the number of data, whereas the simple clustering approach success rate varies between 97% and 85%.

5 Conclusion

In this paper we presented a new multilevel algorithm for clustering web data. It allows to deal with the big mixed databases, such as business, market basket data, which include all type of data in the same row. The algorithm combines a genetic algorithm in the coarsening step, a traditional clustering in the initial solution step, and a supervised classification especially the k-nearest-neighbors algorithm in the refinement step. As our experimental results proved, the best clustering solution is produced by the multilevel clustering approach when comparing to the simple k-means. Furthermore, the algorithm has the additional advantage of being extremely fast, as it is based on a genetic algorithm. For instance, the amount of time required by the proposed algorithm ranges from 53 milliseconds for a web data base with 1000 data, to 2000 milliseconds for a web data base with 45000 data, on a I5 core PC. We believe that this paper presents the first attempt for developing a robust framework for a large scale clustering approaches. However, a number of key questions remains to be addressed, in particular the best way to design the different components of the multi-level paradigm. We can imagine in a future work, an application of the multilevel clustering of large data warehouse, in order to speed up its execution time and also to discover the new knowledge.

References

1. Han, J., Kamber, M.: Data mining: Concepts and techniques. Morgan Kaufmann Publisher (2001)
2. Shih, M.-Y., Jheng, J.W., Lai, L.F.: A Two-Step Method for Clustering Mixed Categroical and Numeric Data. Tamkang Journal of Science and Engineering 13, 11–19 (2010)
3. He, Z., Xu, X., Deng, S.: Clustering Mixed Numeric and Categorical Data: A Cluster Ensemble Approach. CoRR abs/cs/0509011 (2005)
4. Huang, Z.: Clustering large data sets with mixed numeric and categorical values. In: The First Pacific- Asia Conference on Knowledge Discovery and Data Mining (1997)
5. Meta-Knowledge, W., Drias, H., Djenouri, Y.: Multilevel clustering of induction rules for web meta-knowledge. In: Rocha, Á., Correia, A.M., Wilson, T., Stroetmann, K.A. (eds.) Advances in Information Systems and Technologies. AISC, vol. 206, pp. 43–54. Springer, Heidelberg (2013)
6. Czarnul, P., Ciereszko, A., Frązak, M.: Towards efficient parallel image processing on cluster grids using GIMP. In: Bubak, M., van Albada, G.D., Sloot, P.M.A., Dongarra, J. (eds.) ICCS 2004. LNCS, vol. 3037, pp. 451–458. Springer, Heidelberg (2004)
7. Quaresma, P., Rodrigues, I.P.: Cooperative Information Retrieval Dialogues through Clustering. In: Text,Speech and Dialogue, Part-III, pp. 415–420 (2000)
8. Jain, A.K., Dubes, R.C.: Algorithms for clustering data. Prentice Hall, Englewood Cliffs (1988)
9. Álvarez, M., Pan, A., Raposo, J., Bellas, F., Cacheda, F.: Using Clustering and Edit Distance Techniques for Automatic Web Data Extraction. In: Benatallah, B., Casati, F., Georgakopoulos, D., Bartolini, C., Sadiq, W., Godart, C. (eds.) WISE 2007. LNCS, vol. 4831, pp. 212–224. Springer, Heidelberg (2007)
10. Agarwal, P., et al.: International Journal of Engineering Science and Technology (IJEST) 3 (2011) ISSN : 8282-8289

Memory, Experience and Adaptation in Logical Agents

Stefania Costantini and Giovanni De Gasperis

1 Introduction

The DALI language and framework [1] has been exploited in a number of practical applications (cf. [2] for a full list of references on DALI). In order to improve the DALI language and environment, we have been developing a comprehensive approach to memory management.

A first step of our effort has been that of exploring some of the existing approaches to memory management in Psychology and Artificial Intelligence, with special attention to contextual management of memories. In Psychology, the seminal work of Atkinson and Shiffrin [3] proposed a model of human memory which posited two distinct memory stores: short-term memory and long-term memory. Then, [4] introduced the concept of "working memory". In Artificial Intelligence, memory organization and implementation has been treated at some depth in the SOAR architecture (cf. [5] and the references therein) and in the field of design agents [6, 7]. These approaches are widely based upon above-mentioned studies on human memory. Concerning context and contextual reasoning, we can go back to R. Weyhrauch and his FOL system [8]. Since then, the CONTEXT series of workshops has brought new contributions.

In [6, 7], memory is considered to be a "reasoning process", carried out over a suitable representation: in their view in fact, the long-term memory does not simply contain items of information coming from the outside, but also (and probably mainly) agent's beliefs which are the result of the agent's own elaboration of perceptions, experiences and previous internal processing. In [7] it is remarked that memory in an agent is the process of learning or reinforcing a concept. In SOAR [5], beyond short-term memory (graph structure) and long-term memory (production rules) there is an "episodic memory" which contains temporally ordered "snapshots" of working memory and a "semantic memory" which contains symbolic structures

Stefania Costantini · Giovanni De Gasperis
DISIM, L'Aquila, Italy

J. Casillas et al. (Eds.): *Management Intelligent Systems*, AISC 220, pp. 17–24.
DOI: 10.1007/978-3-319-00569-0_3   © Springer International Publishing Switzerland 2013

representing general facts, and provides the ability to store and retrieve declarative facts about the world.

In this paper, we propose a framework for managing memory of agents in DALI, that can be however potentially extended to other agent-oriented languages and formalisms. In this framework, agent memory is created and maintained, and customized according to the present context. This is obtained not by means of rules as usually intended, but by means of special constraints (called A-ILTL axioms [13]) to be dynamically checked with a certain (customizable) frequency. Our approach is complementary rather than alternative to belief revision.

The proposed approach is illustrated in Sections 2– 5, before a concluding discussion in Section 6. We assume that the reader is to some extent familiar with logic programming, and in particular with prolog-like logic programming, and to a small extent with Answer Set Programming (called 'ASP' in the rest of the paper for the sake of conciseness). The reader may refer to [9] for the former and to [10] for the latter, and to the references therein.

2   Advanced Memory Management in Logical Agents: Contexts

Drawing inspiration from the above-illustrated previous approaches, we depict for DALI the following scenario. The semantic memory (say S) in our framework coincides with the agent's initial knowledge base. Short-term memory consists of a set P, where set P plus the Past Constraints plus A-ILTL axioms (see Section 5) constitute the working memory. Long-term memory is set PNV.

The first new element that we intend to introduce is a notion concerning the *context* an agent is presently involved in. The aim is that of customizing working memory contents, and thus the way an agent behaves, according to the present context(s). In this setting, we do not consider a context simply as the location and activity of the agent. More generally, we intend to consider the activity, role and objectives of the agent (and possibly also the location).

For instance, consider an agent that is able to play a number of games (say, e.g., chess, poker, and others, and even gamble on the stock exchange as a particular kind of game). The context here may include the game the agent is presently playing (if any). For enabling the agent to paly, the related competencies need be loaded into the working memory. The context needs not be a single one. Additionally, the agent can be at leisure or at work, at home or somewhere else, and at the same time be either in a normal or in an emergency situation, and so on.

We assume that various agent activities may in principle proceed in parallel. Thus for instance, playing a game and answering the phone may occur in combination.

2.1   Defining Contexts

We can adapt the DALI notion of present events to represent present contexts. We may simply represent contexts by means of a set of (meta-)facts concerning a distinguished predicate *context* (not used for other purposes). The version *contextN* of this predicate will define contexts which are actual *now* for the agent. They will later possibly become (if no longer actual) time-stamped past events with predicate *contextP*. A sample context definition can be the following:

$$context(at\_leisure, at\_work, play\_chess,$$
$$normal, emergency, \ldots).$$
$$contextN(play\_chess).$$
$$contextN(at\_leisure).$$
$$contextN(normal).$$

where the first fact lists all possible contexts, and the following ones the contexts that are active at the moment. More generally, it is required that the agent initial program contains a fact, that we call *context declaration*, of the form:

$context(PC_1, \ldots, PC_r).$

where we call *DC* the set $PC_1, \ldots, PC_r$, and we call the PC_is, which are either constants or ground terms, *contexts*. This declaration states which are the contexts the agent may possibly find itself in, or may choose to set itself in. Facts *contextN* may be initially stated, but this is not strictly required as the agent may want to set a context later. We let *CN* be the set of contexts occurring in facts *contextN(C)* which are present in the knowledge base.

2.2   Set or Change Context

In our setting, the agent can decide to change the context, possibly in reaction to external circumstances but also according to its own preferences about how to proceed. To this aim, the agent can perform a distinguished action (postfix 'A'):

$change\_contextA(M_1, \ldots, M_k; C_1, \ldots, C_n).$
where:
(a) each M_j is an atom and each C_i can be either an atom or a disjunction of atoms in the form: $(C_{i_1} \mid \ldots \mid C_{i_s}) : Prefs$
b) all the M_js, C_i and C_{i_r}s occur in *DC*.

   The part ': *Prefs*' is optional, and expresses in the notation of [11] preferences about which of the C_{i_j}s is preferred under which conditions. The intended meaning is that the agent wishes to switch to contexts $M_1, \ldots, M_k, C_1, \ldots, C_n$ where: (i) the M_js are mandatory, i.e., after the context switch all of them must be present contexts; (ii) the C_is are wished for, i.e., they will become present contexts if possible; (iii) for each C_i which is a disjunction, any of the C_{i_j}s can be selected, either indifferently or according to preferences, if stated. For instance,

$change\_contextA(at\_leisure;$
$$(play\_chess \mid play\_checkers : less\_difficult)).$$
means that the agent intends to switch to a context where it is at leisure and wishfully plays either chess or checkers, preferring the one which is less difficult. Binary predicate $less\_difficult(X, Y)$ must be defined in the agent's knowledge base where whenever $less\_difficult(g_1, g_2)$ holds, $g_1, g_2 \in \{play\_chess, play\_checkers\}$,

the
former is best preferred. This notation extends to sets of elements and to any binary predicate which establishes an ordering over these elements.

A distinguished ASP module is supposed to be defined in order to manage context shift. This module is similar to a "reactive ASP module" as defined in [12]. We adopt an ASP module as it is able to return several solutions, called "answer sets". With a suitable definition of the module, solutions will be compliant to possible complementarity or incompatibilities among contexts: for instance, one cannot be both at leisure and at work, does not play games if in emergency, and, say, one gambles on the stock exchange only when working. The definition of this module is therefore an important part of the definition of an agent. A possible (naive) definition of a module related to our example can be for instance the following (where we remind the reader that rules starting with :- are ASP *constraints* and state that their conditions cannot simultaneously hold, and that an *even cycle* like a :- $not\,b$, b :- $not\,a$ generates indifferently either a or b.

at_leisure :- *not at_work*.
at_work :- *not at_leisure*.
normal :- *not emergency*.
emergency :- *not normal*.
:- *emergency, game*.
game :- *play_chess*.
game :- *play_checkers*.
play_chess :- *not play_checkers*.
play_checkers :- *not play_chess*.
:- *at_leisure, at_work*.
:- *normal, emergency*.

Such a module, that we can call *context-switch module*, may have none, one or several answer sets. We say that a context switch is *enabled* if the context-switch module admits answer sets. If so, we will say that an answer set S of the context-switch module *entails change_contextA*$(M_1, \ldots, M_k; C_1, \ldots, C_n)$ if for every M_j, $j \leq k$, $M_j \in M$, and that S *enables* each atom C_i if $C_i \in S$, and each disjunction C_k if at least one element C_{k_j} of the disjunction is in S.

3   Implementation

The DALI interpreter treats the action of context change $CC = change\_contextA(M_1, \ldots, M_k; C_1, \ldots, C_n)$ as follows, where as said CN is the set of facts of the form $contextN(C)$ included in the knowledge base.

- The context-switch ASP module is invoked, with the M_js as input. I.e., the M_js will be added to the module as new facts.
- If the resulting module has no answer set, then the action fails and a failure past event will be generated, that can possibly be suitably managed by a DALI internal event.

- If the resulting module admits answer sets, than it entails CC by construction. One answer set $\hat{S}$ is selected in the following way.

  - The best preferred answer sets $N_1, \ldots, N_k$ are chosen according to the preferences expressed in the CC elements.
  - Among the N_vs, the answer set $\hat{S}$ is chosen that maximizes the intersection with CN. I.e., as few changes as possible are performed (other strategies are of course possible). If more than one of the N_vs fulfils the requirements, one of them is nondeterministically chosen.

- All facts of the form $contextN(C)$ presently included in the knowledge base are removed, and corresponding facts $contextP(C) : T$ are added to PNV, where T is the present time.
- For each $C \in \hat{S} \cap DC$, a new fact of the form $contextN(C)$ is added to the knowledge base.

4 Context Switching

In the proposed setting, each context is associated to one or more modules (intended as set of rules) aimed at managing the situation the context is about. We assume that such modules are kept in a meta-level format enriching each module with additional information. Let for instance a possible form be the following, where let $\ulcorner M \urcorner$ be any meta-level representation of the module M (that we take here to be a set of DALI rules).

$mod(context(c), name(n), source(a), time(t_1),$
$\quad goal(g), location(l), timeout(t_2),$
$\quad feature(f), eval(v), active(b), lastused(t_3), \ulcorner M \urcorner)$

The elements of the description have the following meaning:

- $context(c)$ indicates which context the module is aimed at managing. c is a constant, might in principle be a list. $name(n)$ specifies the module name.
- $source(a)$ indicates which agent the module has been acquired from. a is the agent's name, can be $self$ if the module is part of the agent program. $time(t_1)$ is the time of acquisition, can be 0 if the module is part of the agent program from the beginning.
- $goal(g)$ is the goal that the module is aimed at reaching. E.g., for a module with context $chess$ the goal can be, e.g., $win\text{-}game$ or $teach\text{-}to\text{-}play$, where $timeout(t_2)$ states the deadline for reaching the goal. The goal may be empty or can be in principle a list, the timeout may be empty. $feature(f)$, if specified (f might be a list), may express some refinement with respect to the context and the goal. For instance, for any game where the goal is to win, the feature might be the level of expertise at which the module is supposed to enable the agent to play.
- $location(l)$ is the location where the agent is supposed to be when pursuing the goal, can be a list of options.
- $eval(v)$ is some kind of rating associated to the module, that should be related to "how good" the module has been in reaching the goal in past usages. The

evaluation can be inherited by sender agent in case of acquired knowledge, and/or can be updated by the agent itself.

- $active(b)$ states whether the module is presently in use or not, if not $lastused(t_3)$ states the last time of module usage.

The set of module descriptions will include none, one or more module(s) for each possible context. In our view, these descriptions are part of the long-term memory. We also enhance the definition of the working memory, that in our setting at this point will be composed not only by the set P plus the A-ILTL axioms, but also by the context declarations, and by one module for each present context. This module will be loaded whenever a context enters into play.

Then, when updating contexts, two more actions have to be performed:

- Eliminate from working memory modules concerning contexts to be removed, and update the corresponding descriptions in the fields *active*, which is set to false, *lastused*, that is set to the time of removal, *eval* that can be updated according to agent's satisfaction (we do not treat this aspect here).
- Load into the working memory one module for each new context. In the extreme case where no module is available, a request might be issued to sibling agents.

It remains to be seen what to do if there is more than one module corresponding to one context. The different modules may correspond to different goals or to the same goal with different features. E.g., for the context *poker* the goals might be either *win-money* or *minimize-loss* and the features might be for instance either *low-risk* or *high-risk*, referring to the style and attitude of the player.

To this aim, we can improve the *change_contextA* format. Precisely, each C_i (or each C_{i_j} for elements of disjunctions can be of the form: $C_i(Goal, Feature)$ e.g., in the above example, *poker(win-money,high-risk)*. The specification of goal and feature is to be intended as optional, but can also be enriched to: $C_i(Goal, Feature\_pref)$. where *Feature_pref* expresses preferences about features, e.g., in the above example, *poker(win-money, low-risk > high-risk* **pref_when** *short-money)*, stating that in the case *short-money* is entailed by the present knowledge base, *low-risk* should be preferred as a feature to *high-risk*, in case both modules are available (otherwise, the choice is indifferent). Another possible choice is whether one might select (given goal and features) the most recent or the best evaluated module.

5 Context Management

A-ILTL meta-axioms for agent runt-time self-checking and self-repair/improvent [13] can be exploited for context-switching. What would it happen, e.g., in case the agent loads a module in the view of a goal, but the goal is not reached within the given deadline? Clearly, the evaluation of the module should be affected accordingly. However, in case alternative modules are available, the agent might wish to remain in the context where it is, but exchange the module "on the fly". For simplicity, let us assume that whenever loading a context C and a related module, an object-level fact is created of the form: $module(C, N, Goal, Location, Timeout, Feature)$

where N is the module name and the other fields (possibly empty) represent goal, location, timeout and features as extracted from the module definition. The above fact is to be removed on removal of the module. Below is a sample A-ILTL axiom that, whenever checked at run-time (at a customizable frequency), replaces a module which has failed its objective with another one (if available).

$$NEVER\,(module(C,N,Goal,Location,Timeout,Feature),$$
$$not\,achieved(Goal),expired(Timeout)) \div$$
$$replace\_module(C,N,Goal,Timeout,Feature)$$

Or, in the case an agent wishes to update the feature, e.g., passing from beginner to expert in some game, let us assume it asserts a fact $new\_feature(C,Goal,F)$. The following A-ILTL axiom would perform a module exchange (if a suitable module is available) whenever checked:

$$NEVER\,(module(C,N,Goal,Location,Timeout,Feature),$$
$$new\_feature(C,Goal,F),F \neq Feature \div update\_module(C,Goal,F)$$

6 Related Work and Discussion

In this paper, we have presented a context-sensitive approach to managing memory and memories in logical agents, born in the context of the DALI language, but adaptable to other logic-programming based agent-oriented languages. To the best of our knowledge the proposed approach is a novelty in the logical agents realm. It drew inspiration from related work in Artificial Intelligence, yet it introduces original aspects, aimed at practical applications of logical agents. It can be interesting to notice that notions of memory and contexts can be usefully exploited in different and seemingly unrelated fields. For instance, [14] discusses how equipping agents with learning abilities, that require both memory and context-sensitivity, often allows for better performance in adaptive networks. As emphasized in [15], several studies have focused on the effect on agent-user interaction (even for robotic agents) of users' mental models about an agent. Namely, the difference between the users' expectations regarding the functions of an agent and the function that they actually perceived would significantly affect the user feedback and degree of acceptance toward the agent. This, the adoption of context-based interactions focused upon user's expectations and goals might potentially improve the interaction effectiveness.

We have conducted a number of experiments to investigate how the use of contexts affects the human users' perception of agents behavior, via the mandatory projects that the students of the "Intelligent Agents" course in L'Aquila have to develop. Students are required to choose a movie or a novel of interest, and implement as DALI agents the main characters, spending particular attention in memory management. A result of this experimentation is the demo presented in [16], where in particular we present an implementation of "Il berretto a sonagli" ("Cap and Bells") by Luigi Pirandello, with DALI agents to "impersonate" the main characters. Aim of future work is to elicit quantitative rather than just qualitative feedback from the experiments.

References

1. Costantini, S., Tocchio, A.: A logic programming language for multi-agent systems. In: Flesca, S., Greco, S., Leone, N., Ianni, G. (eds.) JELIA 2002. LNCS (LNAI), vol. 2424, pp. 1–13. Springer, Heidelberg (2002)
2. Costantini, S.: The DALI agent-oriented logic programming language: References (2012), http://www.di.univaq.it/stefcost/info.htm
3. Pearson, D., Logie, R.H.: Effect of stimulus modality and working memory load on mental synthesis performance. Imagination, Cognition and Personality 23(2-3), 183–191 (2004)
4. Baddeley, A.: Working memory, thought and action. Oxford University Press (2007)
5. Laird, J.E.: Extending the SOAR cognitive architecture. In: Proc. of the First Artificial General Intelligence Conf., pp. 224–235 (2008)
6. Liew, P., Gero, J.S.: Constructive memory for situated design agents. AIEDAM, Artificial Intelligence for Engineering Design, Analysis and Manufacturing 18(2), 163–198 (2004)
7. Gero, J.S., Peng, W.: Understanding behaviors of a constructive memory agent: A markov chain analysis. Knowledge-Based Systems 22(8), 610–621 (2009)
8. Weyhrauch, R.: Prolegomena to a theory of mechanized formal reasoning. Artificial Intelligence 13(1), 133–176 (1980)
9. Apt, K.R., Bol, R.: Logic programming and negation: A survey. The Journal of Logic Programming 19-20, 9–71 (1994)
10. Gelfond, M.: Answer sets. In: Handbook of Knowledge Representation, ch. 7. Elsevier (2007)
11. Costantini, S., Formisano, A., Petturiti, D.: Extending and implementing RASP. Fundamenta Informaticae 105(1-2), 1–33 (2010)
12. Costantini, S.: Answer set modules for logical agents. In: de Moor, O., Gottlob, G., Furche, T., Sellers, A. (eds.) Datalog 2010. LNCS, vol. 6702, pp. 37–58. Springer, Heidelberg (2011)
13. Costantini, S.: Self-checking logical agents. In: Osorio, M., et al. (eds.) Proc. of LA-NMR 2012, CEUR Workshop Proceedings, vol. 911 (2012) Invited paper, CEUR-WS.org
14. Tu, S.Y., Sayed, A.H.: On the influence of informed agents on learning and adaptation over networks. CoRR abs/1203.1524 (2012)
15. Komatsu, T., Kurosawa, R., Yamada, S.: How does the difference between users' expectations and perceptions about a robotic agent affect their behavior? I. J. Social Robotics 4(2), 109–116 (2012)
16. Costantini, S., D'Andrea, A., Gasperis, G.D., Tocchio, A.: DALI logical agents into play. In: Proc. of the AI*IA Workshop Popularize Artificial Intelligence, PAI 2012 (2012)

Mining the Traffic Cloud: Data Analysis and Optimization Strategies for Cloud-Based Cooperative Mobility Management

Jelena Fiosina, Maksims Fiosins, and Jörg P. Müller*

Abstract. Future Internet (FI) technologies can considerably enhance the effectiveness and user friendliness of present cooperative mobility management systems (CMMS), providing considerable economical and social impact. Real-world application scenarios are needed to derive requirements for software architecture and smart functionalities of future-generation CMMS in the context of the Internet of Things (IoT) and cloud technologies. The deployment of IoT technologies can provide future CMMS with huge volumes of real-time data that need to be aggregated, communicated, analysed, and interpreted. In this study, we contend that future service- and cloud-based CMMS can largely benefit from sophisticated data processing capabilities. Therefore, new distributed data mining and optimization techniques need to be developed and applied to support decision-making capabilities of future CMMS. This study presents real-world scenarios of future CMMS applications, and demonstrates the need for next-generation data analysis and optimization strategies based on FI capabilities.

Keywords: Cloud computing architecture, ambient intelligence, distributed data processing and mining, multi-agent systems, distributed decision-making.

1 Introduction

Increasing traffic and frequent congestion on today's roads require innovative solutions for infrastructure and traffic management. As the components of traffic systems

Jelena Fiosina · Maksims Fiosins · Jörg P. Müller
Clausthal University of Technology, Institute of Informatics,
Julius-Albert Str. 4, D-38678, Clausthal-Zellerfeld, Germany
e-mail: {Jelena.Fiosina,Maksims.Fiosins}@gmail.com,
       Joerg.Mueller@tu-clausthal.de

 * The research leading to these results has received funding from the European Union Seventh Framework Programme (FP7/2007-2013) under grant agreement No. PIEF-GA-2010-274881.

J. Casillas et al. (Eds.): *Management Intelligent Systems*, AISC 220, pp. 25–32.
DOI: 10.1007/978-3-319-00569-0_4   © Springer International Publishing Switzerland 2013

become more autonomous and smarter (e.g. vehicles and infrastructure components are now equipped with communication capabilities), there is an increasing need for cooperation among intelligent transportation systems (ITS) for traffic management and environmental monitoring in order to improve traffic management strategies. Further, there is growing interest and increasing volume of investments to cooperative mobility management systems (CMMS). In these new-generation business management systems, the management of transportation networks is closely integrated with the business strategies and operational models of transport companies and individual customers, providing a considerable impact for companies in terms of business planning, service quality and adaption to customer needs as well as for individual users in terms of time and money saving, adaptive travel planning and support of social mutually beneficial behavior. Innovative cloud services can be created using the cloud capabilities of future Internet (FI) to access smart objects via the Internet of Things (IoT). This development can enable wide access to necessary information, because all of this data will be available in-the-cloud.

However, implementing a traffic cloud is far from easy. From an end user's point of view, the complexity of data and algorithms is hidden in the cloud. Users(ranging from traffic authorities to car drivers and automated components) expect to work with relatively simple applications on the Internet via mobile or embedded devices. These devices are fully connected and can (theoretically) use all the information available from all other users and system elements. This creates great opportunities for coordinated near-optimal management of the infrastructure (e.g. in terms of load balancing). However, there is a huge amount of available data with a short update rate. This creates a need to employ innovative data mining and corresponding decision-making algorithms (under the hood of the traffic cloud) to support CMMS applications in finding, collecting, aggregating, processing, and analysing information necessary for optimal decision-making user behavior strategies. Note that information here is virtually centralized by cloud technologies. However it is distributed, and (very often) created and managed in a decentralized fashion on the physical (fabric) layer. Thus, data mining and decision-making methods are required to find an optimal balance between decentralized information processing/decisions and costs of data transfer/decision coordination.

The contribution of this study is fourfold: First, we analyse related cloud-based architectures and CMMS scenarios. Second, we consider architectures for implementing the corresponding data analysis and optimization of mobility operations. Third, we discuss the employment of appropriate mathematical methods for three use-cases; fourth, we point out and discuss work directions and opportunities in the area of cloud-enabled CMMS.

The remainder of this paper is organized as follows. Section 2 reviews related work in the area of FI and cloud architecture for mobility application. In Section 3, we propose and analyse three application scenarios of CMMS and consider data analysis and optimization of participants' behaviour strategies in traffic systems. In Section 4, we present a cloud-based architecture for mobility networks based on the previously presented scenarios. Section 5 concludes and discusses future research opportunities.

2 Related Work

A strong worldwide interest in opportunities in transportation and mobility field has spurred the need for further analysing these FI opportunities. In Europe, FI and IoT research has been a priority direction for the 7th European Framework Programme (FP7) and will continue to do so for the upcoming Horizon 2020 Programme (e.g. the objectives 'A reliable, smart and secure IoT for Smart Cities' or 'Integrated personal mobility for smart cities' in FP7 or 'Substantial improvements in the mobility of people and freight' in Horizon 2020). These research questions are motivated and co-funded by private companies and municipalities from the areas of transport, logistics, communication and traffic management (e.g. the FP7 project Instant Mobility [1]). These stakeholders understand the possible enhancements to existing systems that new technologies can provide to CMMS. Research in this area is still largely at the stage of formulation of scenarios and coordination protocols. The first cloud-based traffic network architectures have been proposed in [7], which employ ambient intelligence (AmI) [8] or IoT components [6], [7].

An architecture of AmI-enabled CMMS is proposed in [8]. It supports virtual control strategists and management policy makers in decision-making and is modelled using the metaphor of autonomous agents. AmI is defined as the ability of the environment to sense, adapt, and respond to actions of persons and objects that inhabit its vicinity. Moreover, the multiagent system (MAS) paradigm makes AmI environments act autonomously and socially, featuring collaboration, cooperation, and even competitive abilities.

Cloud computing systems are oriented towards a high level of interaction with their users, real-time execution of a large number of applications, and dynamic provisioning of on-demand services. In this study, we consider the layered architecture of cloud-based computing systems presented in [6]. It supports a class of specialized distributed systems that is characterized by a high level of scalability, service encapsulation, dynamic configuration, and delivery on demand. The architecture includes the following layers:

The **fabric layer** includes all computing, storage, data, and network resources available in the cloud. The resources are accessible through the resource services, are used for cloud computations, management, and as testbeds. The **unified source layer** provides infrastructure-as-a-service by defining unified access to the raw computational resources of the fabric layer using a virtual machine. The **platform layer** provides platform-as-a-service, including a collection of specialized tools, middleware, and services on the top of unified resources to create a deployment platform (e.g. scheduling create service and artificial testbeds). The **application layer** contains all applications that are run in the cloud. Application execution in the cloud is distributed: applications can be partly executed on the client, partly in the cloud.

The application of cloud-based architectures for ITS is demonstrated in [7]. In order to provide an acceptable level of service, a cloud-based ITS consists of two main components: an *application component*, which provides dynamic services and runs all the cloud applications; and a *digital (simulated) traffic network* component, which performs constant information collection and processing in order to provide in-time data. A cloud-based ITS adapts its decisions by using available information and by interacting with human as well as automated traffic participants.

We apply our experience in implementing data processing, mining [3], [2], and decision-making methods [4], [5] for existing transportation problems. Next, we discuss the key aspects of methods in future-generation CMMS.

3 Traffic Cloud Scenarios and Related Data Analysis and Optimization Problems

We propose three cloud-based ITS application scenarios: 1) **A cooperative intersection control**, which optimizes vehicle flows in traffic networks by regulating the intersection controllers. 2) **A personal travel companion**, which provides dynamic planning and monitoring of multimodal journeys to travelers, surface vehicle drivers, and transports operators. 3) **A logistics services companion**, which provides benefits to clients and stakeholders involved in, affected by, or dependent on the transportation of goods in urban environments. We demonstrate the most important stages of data processing and optimization in order to derive requirements for a general architecture described in the next section.

3.1 Virtualized Cooperative Intersection Control

This scenario uses adaptive, semi-distributed traffic management strategies hosted in the cloud for the regulation of intersection controllers, and creates ad-hoc networks in the cloud between clusters of vehicles and the traffic management infrastructure. It recommends the optimal speed to drivers to keep the traffic flow smooth, and assists adapting traffic controllers (e.g. traffic lights, signs) based on the real-time traffic situation. This service uses real-time traffic information and a route-data collection service to formulate strategies for the optimization of network operation.

Stage 1: Processing the following data streams (historical and real-time): 1) floating-car data (speeds, positions, etc.); 2) sensor data from the infrastructure (loops, traffic lights, etc.); 3) information about routes and actual locations of collective transport (public transport, taxi, shared cars, etc.) 4) data from distribution vehicles (logistic transport); 5) weather conditions; 6) accidents, car breakdowns, road-works; 7) organizational activities (sport events, conferences, etc.)

Stage 2: Creating ad-hoc networks, which are virtual abstract networks for solving specific problems (intersection and regional traffic models, green wave models, public transport priority, jam avoidance, etc.). Estimating network parameters (traffic flux, density, and speed, travel time estimation, etc.).

Stage 3: Developing static strategies of intersection control and cooperation based on historical information, previous experience, and data models from the previous stage (traffic light signal plan optimization; signal plans for expected events (such as increase of flows, changed weather conditions, organizational activities); cooperation plans of clusters of vehicles, etc.).

Stage 4: Combining dynamic real-time information with static strategies in order to receive up-to-date controlling decisions (correction of signal plans according to current conditions, cooperation of signal controllers to resolve problems such as jams, accidents, etc.)

3.2 Dynamic Multi-modal Journey Planning

The purpose of this use case is to help travellers plan and adjust a multi-modal, door-to-door journey in real-time. It provides improved (i.e., quicker, more comfortable, cheaper, and greener) mobility to daily commuters and other travellers by identifying optimal transportation means and a strong real-time orientation. This planning proposal for a multi-modal journey takes into account the current means of transportation, the traveller's context and preferences, city traffic rules, and the current requirements and constraints. The journey plan needs to obtain an overall indication of the trip duration as well as accommodate early reservation of resources (train or plane ticket).

Stage 1: Processing of the following data streams (historical and dynamic) in addition to the previous application: 1) floating passenger data; 2) travellers' preferences; 3) timetables and availability of collective transport (tickets, shared cars availability, etc.); 4) changes in time-tables.

Stage 2: Creation of ad-hoc networks (transit stations, public transport coordination, passenger choice of transport, etc.) and estimation of network parameters (travel time for different transport modes depending on various factors, waiting times, passenger arrival at stops, price models, etc.).

Stage 3: Multi-modal route pre-planning based on historical data and estimated network parameters for expected conditions (pre-planning for popular routes, pre-planning for pre-booked routes, pre-planning for expected events) as well as optimal time-table calculation for public transport based on the expected conditions.

Stage 4: Dynamic update of pre-planned routes for the actual multi-modal journey (actual travel-time estimation, re-planning in the case of delays in previous trips in the multi-modal chain, re-planning for additional travel possibilities, or cancelling a part of the multi-modal journey), as well as dynamic update of public transport time-tables (on-demand changes, co-ordination of different transport means).

3.3 Itinerary Booking and Real-Time Optimized Route Navigation

This use case helps a logistics provider (1) guarantee quick (especially on-time) deliveries at a low cost based on up-to-date information and (2) maximize the efficiency of each vehicle and the fleet. It is fundamental to optimize the movements of the logistics vehicles, to help them avoid traffic jams and take the shortest routes when possible.

Stage 1: Processing of the following data streams (historical and dynamic) in addition to the first application: 1) order data (transportation demand); 2) available logistic vehicles (possible load, speed, etc.); 3) timetables (if necessary) and actual positions of the vehicles; 4) client data (drop-off preferences, actual location, etc.).

Stage 2: Creation of ad-hoc networks (delivery models, logistic provider-client interaction models, etc.), and estimation of the network parameters (travel times for different route segments, delay probability, drop-off process time distribution, probability of accidents, probability of problems with vehicles, etc.).

Stage 3: Pre-planning of the delivery process (preliminary good distribution by vehicles, preliminary order of clients for each vehicle, preliminary route for each vehicle, preliminary time window for each client, etc.). Note that the itineraries of large logistic operators can be used to provide better predictions of the traffic situation using virtualized cooperative intersection intelligence application as well as by applying priority rules for logistic vehicles during booking.

Stage 4: Dynamic update of pre-planned delivery routes depending on up-to-date information (re-planning of routes depending on current traffic situation, re-planning in the case of accidents or traffic jams, re-planning in the case of vehicle problems, estimation of actual delivery time, etc.). Cooperation between logistic vehicles (exchange or orders, adoption of other vehicle's orders in the case of problems, etc.). Dynamic agreement with clients (agreement about drop-off place depending on current position of the vehicle and client, agreement about change of drop-off time, reaction to the new/changed customer requests, etc.).

4   Reference Architecture for Traffic Cloud Data Mining and Strategy Optimization

The applications mentioned in the previous section are data-intensive. Services provided through the cloud require large amounts of data to be processed, aggregated, and analysed. Then, the processed data is used for calculating optimal strategies for traffic participants. Now we generalize the stages of data processing and network optimization from the scenarios discussed in the previous section. We propose a reference architecture for traffic cloud data mining and optimization of strategies (TCDMOS), which is based on [7], but we focus on data processing and decisions. TCDMOS is illustrated in Fig. 1. It includes the following stages of data processing and network optimization:

Stage 1: *Mining data from the IoT and its pre-processing.* All the participants of the cloud-based system have virtual representations as active IoT components (agents). These virtual agents are associated with data (mostly real-time) and act as data sources for the cloud-based system. The cloud system locates and collects the necessary data from different agents, and provides usual data mining operations (changes and outliers are detected, preliminary aggregation and dimensionality reduction are performed). The collected data are stored as historical information in the cloud and are used later as input data for ad-hoc network models (Stage 2). Stream-based

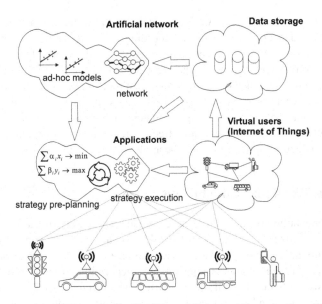

Fig. 1 TCDMOS Architecture: Traffic Cloud Data Mining and Optimization of Strategies

methods of semi-decentralized change-point detection, outlier detection, clustering and classification, and factor analysis occur regularly in this stage.

Stage 2: *Ad-hoc network models.* The application-specific digital networks of virtual traffic participants (e.g. regional, social) are created, and the corresponding data models are used in order to estimate the important characteristics and parameters of these networks using the information collected in Stage 1 and for strategy optimization at Stage 3. The future behaviour of traffic participants is forecasted as well. Semi-decentralized, flows forecasting (possibly with incomplete information) methods such as (multiple-response) regression models, Bayesian networks, time series, simulation, are also applied at this stage. Many pre-defined data models can run concurrently in the digital network. The corresponding data storages are located in the cloud and are semi-centralized, so the methods should take costs of different pieces of information into account.

Stage 3: *Static decisions and initial strategy optimization.* Cloud applications use pre-calculated results of the ad-hoc network models from Stage 2 and the available historical information (including private information) about the traffic network to perform their pre-planning tasks. Initial optimization of the strategies is resource-expensive, and can be partially pre-calculated in ad-hoc network models and then instantiated according to the application's goals and preferences. These models are also checked in the digital traffic network. This stage can require aggregation of different data models and existing strategies. Methods of self-learning stochastic (multi-criteria) optimization such as neural networks, decision trees, Markov decision processes, choice models, graph optimization algorithms are used.

Stage 4: *Dynamic decisions and strategy update.* The pre-planned tasks from Stage 3 are executed, and updates are made according to the dynamic real-time situation extracted from the virtual agents. The aggregation of the pre-planned data and strategies with the dynamic ones is the most important problem at this stage. An additional difficulty here is the requirement of fast real-time execution. (Automatic) cooperation between users in their decisions is possible; therefore, stream-based methods of data models and strategy updates such as reinforcement learning, Bayesian networks, dynamic decision trees, stream regression, and distributed constraint satisfaction/optimization can be applied.

5 Future Work and Conclusions

The main contribution of this study is a reference architecture for traffic cloud data mining and optimization of strategies (TCDMOS) and related data processing and network optimization methods. We envisage this as an important step towards making FI and cloud technologies usable for next-generation CMMS. TCDMOS requirements were elicited from traffic scenarios, which reflect needs and impact of CMMS for business and society, and the corresponding problems that should be solved for effective cloud system operation were illustrated. Future work will be devoted to elaborating the architecture, developing novel algorithms, and integrating and validating them in state-of-the-art cloud computing frameworks.

References

1. 7-th european framework programme project, instant mobility: Multimodality for people and goods in urban area, cp 284806, http://instant-mobility.com/
2. Fiosina, J.: Decentralised regression model for intelligent forecasting in multi-agent traffic networks. In: Omatu, S., Paz Santana, J.F., González, S.R., Molina, J.M., Bernardos, A.M., Rodríguez, J.M.C. (eds.) Distributed Computing and Artificial Intelligence. AISC, vol. 151, pp. 255–264. Springer, Heidelberg (2012)
3. Fiosina, J., Fiosins, M.: Distributed cooperative kernel-based forecasting in decentralized multi-agent systems for urban traffic networks. In: Proc. of Ubiquitous Data Mining (UDM) Workshop of ECAI 2012, Montpellier, France, pp. 3–7 (2012)
4. Fiosins, M., Fiosina, J., Müller, J.P.: Change point analysis for intelligent agents in city traffic. In: Cao, L., Bazzan, A.L.C., Symeonidis, A.L., Gorodetsky, V.I., Weiss, G., Yu, P.S. (eds.) ADMI 2011. LNCS, vol. 7103, pp. 195–210. Springer, Heidelberg (2012)
5. Fiosins, M., Fiosina, J., Müller, J.P., Görmer, J.: Reconciling strategic and tactical decision making in agent-oriented simulation of vehicles in urban traffic. In: Proc. of 4th International ICST Conference on Simulation Tools and Techniques, SimuTools 2011 (2011)
6. Foster, I.: Cloud computing and grid computing 360-degree compared. In: Proc. of the Grid Computing Environments Workshop, pp. 1–10 (2008)
7. Li, Z., Chen, C., Wang, K.: Cloud computing for agent-based urban transportation systems. IEEE Intelligent Systems 26(1), 73–79 (2011)
8. Passos, L., Rossetti, R., Oliveira, E.: Ambient-centred intelligent traffic control and management. In: Proc. of the 13th Int. IEEE Annual Conf. on ITS, pp. 224–229 (2010)

Towards a Knowledge-Driven Application Supporting Entrepreneurs Decision-Making in an Uncertain Environment

Gianfranco Giulioni, Edgardo Bucciarelli, Marcello Silvestri, and Paola D'Orazio

Abstract. In this paper we present an interactive web-based application which could allow to explore entrepreneurial management. The main topic of the investigation is the analysis of entrepreneurship as a judgmental decision making under uncertainty. At this stage we outline the theoretical background which is behind the software and its basic functioning. We maintain that the fundamental research, which underpins our model, could stimulate new way of thinking to the development of a knowledge-driven tool in supporting risk-taking propensity.

1 Introduction

Recently, the field of mobile decision support system has receiving an increasing interest among managers. The development of information technology and the computational power of modern computer have undoubtedly supported and improved organizational decision making activities (for a literature review see [8]).

Although a clearly definition of decision support system (DSS) is limited to a broad range of conceptual perspectives, we can refer it as any types of information systems that support decision making [10, 21].

According to [18] a DSS can be extended to five categories: communications-driven, data-driven, document driven, knowledge-driven and model-driven. This taxonomy is important in defining specific scopes which are behind the building of an appropriate interactive computer-based system.

However, it can be said that a consensus has been reached on the architecture a DSS must involve: the database, the model, and the user interface [20].

Gianfranco Giulioni · Edgardo Bucciarelli · Marcello Silvestri · Paola D'Orazio
Department of Economic-Quantitative and Philosophical-Educative Sciences,
Viale Pindaro 42 - 65127, Pescara, Italy
e-mail: {g.giulioni,e.bucciarelli,paola.dorazio}@unich.it,
    marcello.silvestri@yahoo.it

J. Casillas et al. (Eds.): *Management Intelligent Systems*, AISC 220, pp. 33–39.
DOI: 10.1007/978-3-319-00569-0_5   © Springer International Publishing Switzerland 2013

Within this orientation we present our theoretical knowledge-driven DSS which in turn is rooted in modeling a specific set of business decisions. We refer to a theoretical model since, at this early stage, it deals with a fundamental economics research on decision making under risk and uncertainty (on this topic see [1]).

Nevertheless, we maintain that our interactive web-based application could support organizations in searching for a way to develop a practical application in supporting risk management activities.

The paper is structured as follow. In section 2 we discuss the theoretical issue which is behind the implementation of the web application we propose. Section 3 shows the functioning of the interactive web-based application. In section 4 we reflect on some managerial implications of our contribution. Section 5 draws the conclusions.

2 Theoretical Basis

The standard methodology of Economics considers the entrepreneur as a fully rational profit-maximizing agent [16, 13].

A widespread criticism towards the traditional theory of firms has been arising in the last thirty years. Several economists actually claim for a multi-disciplinary managerial and behavioral theory of the firm (see, among others, [7]).

The field of organizations and strategic management has been one of the most prolific of academic investigation, but very little has been made to integrate entrepreneurship into the theory of the firm.

According to Casson [5] this could be the consequence of the traditional indifference to firms' management and its psychological assumptions. The conceptualization of entrepreneurship as an institutional aspect of the theory of the firm is therefore regularly neglected [6].

We shed some light on this issue following Hyman Minsky's approach [15]. In Minsky's economic thought the analysis of entrepreneurship is related to two interdependent aspects. On one hand entrepreneurs and/or managers determine which assets are to be held. On the other they decide how to finance the ownership or the control of these assets. In a capitalistic economy, the management of the firm's balance sheet structure is the fundamental *speculative decision* which:

> centers around how much, of the anticipated cash flow from normal operations, a firm, household, or financial institutions pledge for the payment of interest and principal on liabilities [14, p. 84].

It is the *portfolio choice* - that moves through real calendar time and fundamental uncertainty - which underpins financial behavior. Minsky pointed out that firms' patrimonial structure (their balance sheet) affects their costs. Then, an analysis of the financial aspects should be added to the traditional microeconomic framework in order to understand firms' behavior in a risky and uncertain environment.

Our contribution to the entrepreneurship research in economic and managerial theories is deeply rooted in the Hyman Minsky's economic thought which we refer to induce his theory of the firm by the building of an interactive web-based application. Under a subjective perception of risk, subjects have to manage the financial structure of their virtual firms.

3 The Interactive Web-Based Application

In this section we briefly explain the functioning of the interactive web-based application.[1] Using Java programming language, we built a graphical user interface which gives the subjects the opportunity to test their intuition and ability as managers. They have to lead a production firm having a simple balance sheet: production is the only item in the assets section, and debt and equity are the items in the liabilities section. The table below shows an example:

Assets	Debt	Equity
10000	9000	1000

In each period the subjects have the following information service.

The report on demand. In this report the subjects can observe a forecasting service of the demand they had in the three past periods and the demand forecast for the current period; the demand obtained by their virtual firm in the three past periods; the production choices the subjects made in the three past periods; the gaps between demand and production in these periods.

The report on patrimonial and economic accounting. This report displays the data of firms balance sheet (assets, debt and equity) and the percentage of assets financed by debt; the data on their performance (the obtained profit, the maximum achievable profit, the ROI of the current period, the average ROI, the number of bailouts and their score). These values are updated after each decision.

The report on charts. These charts show the whole history of the demand (green line) and of the choice of production made by the subjects (red line); the percentage of debt (debt / liabilities*100); the Return on Investment (ROI = profit (or loss) / assets*100).

The information service is displayed in the computer screen by the interactive web-based application reported hereafter.

In each period, subjects are asked to make real and financial choices. A real choice is to choose the production (subjects goal is to rightly guess the demand which will be received by their firm). This choice modifies the levels of assets and debt in the balance sheet (equity does not change). The production capacity can be adjusted by changing assets in the balance sheet (subjects get one product per unit of assets: in the example of the table, firm makes 10000 products).

---

[1] For a detailed presentation see [9].

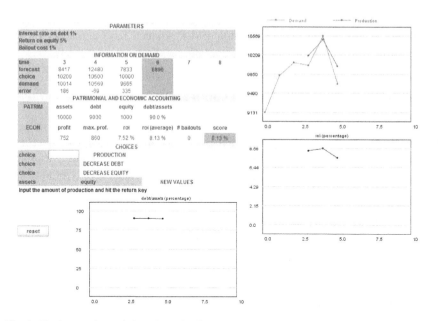

Fig. 1 The interactive web-based application

Once subjects have made their choice of production, the demand is revealed and
the economic result (profit or loss) is computed as the difference between revenue,
production costs and financing costs. Financing costs are as follows: the interest rate
on debt is 1 percentage; the cost of equity is 5 percentage.

The information on the maximum achievable profit is given beside the economic
result realized in the current period. Given the financial structure, the maximum
profit is achieved when the demand is equal to production: the larger is the gap
the lower is subjects economic result. Subjects can suffer a loss in either cases of
excesses and shortages of production.

Finally, the ROI of the current period, the average ROI up to the current period,
the number of bailouts and the score (average ROI minus bailouts) are updated after
the production choice is made.

Regarding financial choices subjects can have different scenario.

Subjects have a profit. They have to choose any amount of debt to be repaid to the
bank. They cannot repay an amount higher than the profit of the current period. If
they choose a positive amount, debt and equity are modified, but assets remain unchanged in the balance sheet. Profits unused to reduce debt are subjects management
virtual reward.

Subjects have a loss higher than equity. A loss reduces assets in the balance sheet,
and if the loss is higher than equity subjects must bailout their firm. The bailout
procedure asks subjects to reset their balance sheet: new levels of assets and equity
have to be chosen. In the case of bailout, a penalty of 1 percentage is charged to
their average ROI (each bailout reduces subject score by 1 percentage).

Subjects have a loss lower than equity or a profit. They have to choose to reduce the amount of capital if it is deemed too high. If a positive value is chosen, debt and equity are modified, but assets remain unchanged in the balance sheet.

During the simulation subjects goals are to limit the number of bailouts and to maximize the performance of the firm. The return on investment in percentage (ROI = profit(loss) / assets*100) is taken as measure of performance. These two aspects are summarized by the score which is calculated as average ROI minus number of bailouts. Subjects reward as a participant to the test depends on their score.

High scores can be achieved by setting production as close as possible to the demand which will be observed on the market; and by taking the appropriate decisions regarding the financial structure of the firm. Indeed subjects have to choice a trade off between profitabilty/fragility. The equity is a cushion of safety against loss but at the same time it is a source of financing more expensive than debt (i.e. see pecking order theory [17].

4 Towards a Knowledge-Driven DSS

The interactive web-based system introduced in the previous section holds many potentialities for innovative techniques especially in the field of knowledge, education and training management. The field of corporate e-learning is facing a big challenge and it concerns the need to design, develop and deliver new solutions to entrepreneurial management [3].

In order to improve or implement training strategies firms can make workplaces more conducive to learning and can promote environments in which knowing and learning are constructed through ongoing and reciprocal processes.

In Steven Billets terms, there is a need for more empirical studies in order to get a better understanding of workplace learning and training in specific business areas, including some resorurces and dynamic capabilities that are buried deep within everyday workplace practices and conditions [2].

The focus we propose in this paper is on the getting of production and financial scenarios and their simulations which involves the necessity of dealing with continuous improvements in management training, new skills and new practice. Many studies show the strong statistical significance of those variables concerning learning nature of the subject environments, whereby students, employees and managers have participated in improvement groups or have submitted advanced suggestions, have been interviewed for performance evaluation purposes, and are involved and consulted by the firm (among others, see [11, 4]).

To this aim we design an interactive web-based application with which subjects have to interact. We give subjects the opportunity to learn by doing in a management situation which is a simplified simulated experience of the real business world.

Workplaces can experience different types of learning (formal and/or informal). As regarding informal learning, the web application presented in this paper is based on an innovative experimental economics technique allowing us to explore a number of economic and financial aspects of the entrepreneurial management.

We maintain that Experimental Economics [12] could be a research area of interest for knowledge-driven DSS. This kind of decision support system is usually designed in order to suggest or recommend actions to managers. On the other hand, Experimental Economics is the application of experimental methods to study economic questions [19]. By the building of a controlled architecture (or environment) it is possible to gather all the relevant data. Nevertheless it is possible to improve this architecture, through workplaces experiences, in searching for a practical application which can support risk-taking propensity.

5 Conclusions

This paper suggests a new way to overcome theoretical problems of the traditional theory of the firm which has proven to be not adequate to deal with managerial and behavioral assumptions.

Our proposal is deeply rooted in Minsky's economic thought which we used to frame an artificial economic environment and to build an interactive web-based application.

The key device is the Java programming language. As it is well known, each web browser is (or can be easily) endowed with the capability to download and run Java programs (called Applets) simply pointing to an URL. So, by using the Java language, it is possible to build a Graphical User Interface showing all the relevant data to take a decision (labels, charts of historical data and so on) and text filed where subject input his/her choice. By using the Java Database connectivity and the Java driver for the chosen database management system, subject decisions and the information describing the settings in which they was taken can be recorded in a remote database server (as an example for MySQL, which is the most popular freely available database management system).

In this way it is possible to design an environment, store knowledge and make it available (i.e. data mining). The prototype developed here, although refers to a fundamental research, could generate new knowledge to practitioners for the building of a more reliable risk management decision support system. Nevertheless the web application could be particularly useful for education management, recruitment and selection as well as for training/e-learning.

References

1. Abdellaoui, M., Hey, J.D.: Advances in decision making under risk and uncertainty. Springer, Berlin (2008)
2. Billet, S.: Learning through practice. Springer, The Netherlands (2010)
3. Brynjolfsson, E., Hitt, L.M.: Beyond computation: Information technology, organizational transformation and business performance. Journal of Economic Perspective 14, 23–48 (2000)
4. Caroli, E., John, J.V.R.: Skill-biased organizational change? evidence from a panel of british and french establishments. The Quarterly Journal of Economics 116, 1449–1492 (2001)

5. Casson, M.: Entrepreneurship and the Theory of the Firm. Journal of Economic Behaviour and Organization 58, 327–348 (2005)
6. Ebner, A.: Entrepreneurship and Economic Development. Journal of Economic Studies 32, 256–274 (2005)
7. Foss, N.J.: Resources, Firms and Strategies: A Reader in the Resource-Based Theory of the Firm. Oxford University Press, London (1997)
8. Gao, S.: Mobile decision support system research: a literature analysis. Journal of Decision Systems 22, 10–27 (2013)
9. Giulioni, G., Bucciarelli, E., Silvestri, M.: A model implementation to investigate firms financial decisions (2011),
   www.dmqte.unich.it/users/giulioni/model_description.pdf
10. Holsapple, C.W., Whinston, A.B.: Decision Support Systems: a Knowledge based Approach. West Publishin Co., Minneapolis (1999)
11. Ichniowski, C., Kathryn, S., Prennushi, G.: The effects of human resource management practices on productivity: A study of steel finishing lines. American Economic Review 87, 291–313 (1997)
12. Kagel, J.H., Roth, A.: Handbook of Experimental Economics. Princeton University Press, Princeton (1995)
13. Kydland, F.E., Prescott, E.C.: Time to Build and Aggregate Fluctuations. Econometrica 50, 1345–1370 (1982)
14. Minsky, H.P.: Stabilizing an Unstable Economy. Yale University Press, New Haven (1986)
15. Minsky, H.P.: Induced Investment and Business Cycle. Edward Elgar Publishing, Northampton (2004)
16. Muth, J.F.: Rational Expectations and the Theory of Price Movements. Econometrica 29, 315–335 (1961)
17. Myers, S.C., Majluf, N.S.: Corporate financing and investment decisions when firms have information that investors do not have. Journal of Financial Economic 13, 187–221 (1984)
18. Power, D.: Decision Support Systems: Concepts and Resources for Managers. Quorum Books, Westport (2002)
19. Smith, V.L.: Method in experiment: Rhetoric and reality. Experimental Economics 5, 91–110 (2002), doi:10.1023/A:1020330820698,
    http://dx.doi.org/10.1023/A:1020330820698
20. Sprague, R.H., Carlson, E.D.: Building Effective Decision Support Systems. Prentice-Hall, Englewood Cliffs (1982)
21. Sprague, R.H., Watson, H.J.: Decision Support for Management. Prentice-Hall, Englewood Cliffs (1996)

Training Neural Networks by Resilient Backpropagation Algorithm for Tourism Forecasting

Paula Odete Fernandes, João Paulo Teixeira, João Ferreira, and Susana Azevedo

Abstract. The main objective of this study is to presents a set of models for tourism destinations competitiveness, using the Artificial Neural Networks (ANN) methodology. The time series of two regions (North and Centre of Portugal) has used to predict the tourism demand. The prediction for two years ahead gives a mean absolute percentage error between 5 and 9 %. Therefore, the ANN model is adequate for modelling and prediction of the reference time series. This model is an important and useful framework for better planning and development of these two regions as they operate in highly competitive markets.

Keywords: Neural Networks; Time Series Analysis; Tourism Demand Forecasting; Resilient Backpropagation Algorithm.

1   Introduction

The review paper about techniques for forecasting tourism demand, Witt and Witt [22] offer a viewpoint on the several methods used for forecasting tourism demand, including a comparison of the accuracy of the methods.

More recently, Artificial Neural Networks (ANN) has been used in tourism demand researches (e.g. [15][17][16][6]). The outcomes of these studies are encouraging when compared with other models such as ARIMA [6], linear regression model [14], multivariate adaptive regression splines [13] and multiple regression, exponential smoothing, moving average and naïve [11]. Consequently academics and practitioners continue to improve and develop models and methods

Paula Odete Fernandes · João Teixeira
Polytechnic Institute of Bragança, Apartado 134, 5301-857 Bragança, Portugal
e-mail: {pof,joaopt}@ipb.pt

Paula Odete Fernandes · João Ferreira · Susana Azevedo
NECE - Research Unit, University of Beira Interior, Covilhã, Portugal

João Ferreira · Susana Azevedo
University of Beira Interior, Pólo IV - Edifício Ernesto Cruz, 6200-209 Covilhã, Portugal
e-mail: {jjmf,sazavedo}@ubi.pt

J. Casillas et al. (Eds.): *Management Intelligent Systems*, AISC 220, pp. 41–49.
DOI: 10.1007/978-3-319-00569-0_6   © Springer International Publishing Switzerland 2013

to bring about greater understanding of the economics and business principles as guidance for more effective management and planning in the tourism sector. This artificial intelligence tool can be used as a planning tool for management in the hospitality and catering sectors such as in a higher level, namely the investment decision in the tourism sector.

This paper designs a new neural network model for tourism forecasting which uses the Resilient Backpropagation Algorithm for two tourism destinations in Portugal - The North and Central regions.

2   Artificial Neuronal Network Methodology

Neural networks are the most versatile nonlinear models that can represent both nonseasonal and seasonal time series. The most important capability of neural networks compared to other nonlinear models is their flexibility in modelling any type of nonlinear pattern without the prior assumption of the underlying data generating process [7].

A neural network is composed of a number of interconnected artificial neurons, nodes, perceptrons or a group of processing units, which process and transmit information through activation functions. The connections between processing units are known as synapses and every connection in a neural network has a weight attached to it. The functions most frequently used are the linear and the sigmoidal functions - the logistic and hyperbolic tangent functions. It should also be mentioned that the neurons of a network are structured in distinct layers (better known as the input layer, the intermediate or hidden layer and the output layer), with the ones most commonly used for the forecasting of time series being the multi-layers [2], so that a neuron from one layer is connected to the neurons of the next layer to which it can send information [1].

There are countless learning methods for neural networks. However, they can be classified into two groups, namely supervised and unsupervised method. Supervised learning requires historical data with examples of both dependent and independent variables to train the network. The known answers are worked as a teacher to correct the behaviour of the training network. Unsupervised learning method creates its own model to interpret the data without known answers [8][1][3][16]. Much of the current interest in the neural network technique forecasting tool can be traced to the development of the learning algorithm. The most frequently used algorithm in the forecasting of time series is the backpropagation algorithm, a technique first developed by Werbos [21] and further advanced by Rumelhart and McCleland [19]. This algorithm gives the network the ability to form and modify its own interconnections in a way that often rapidly approaches a goodness-of-fit optimum. Many variants of the Backpropagation training algorithm were developed. Usually, the learning process involves the following stages [23][7][10][3]:

1. Assign random numbers to the weights;
2. For every element in the training set, calculate output using the summation functions embedded in the nodes;
3. Compare computed output with observed values;
4. Adjust the weights and repeat steps (2) and (3) if the result from step (3) is not less than a threshold value; alternatively, this cycle can be stopped early by reaching a predefined number of iterations, or the performance in a validation set does not improve.
5. Repeat the above steps for other elements in the training set.

For an ANN model the prediction equation for computing a forecast of Y_t, using selected past observations, can be written as [3]:

$$Y_t = b_{2,1} + \sum_{j=1}^{n} \alpha_j f\left(\sum_{i=1}^{m} \beta_{ij} y_{t-i} + b_{1,j} \right) \tag{1}$$

where,

m , number of nodes in the input layer;

n , number of nodes in the hidden layer;

f , sigmoidal activation function;

$\{\alpha_j, j = 0,1,...,n\}$, vector of weights that connects the nodes of the hidden layer to those of the output layer;

$\{\beta_{ij}, i = 0,1,...,m; j = 1,2,...,n\}$, weights that connect the nodes of the input layer to those of the hidden layer;

$b_{2,1}$ and $b_{1,j}$, indicate the weights of the independent terms (bias) associated with each node of the output layer and the hidden layer, respectively.

The equation also indicates the use of a linear activation function in the output layer.

3   Artificial Tourism Demand Modelling and Forecasting

3.1   Designing Artificial Neural Networks Model

For the selection of data we used the secondary source published in the Portuguese National Statistics Institute. The time series used in this study for forecasting was "Monthly Guest Nights in Hotels in the North" [GRN] and "Monthly Guest Nights in Hotels in the Centre" [GRC] of Portugal, registered between January 1987 and December 2009, corresponding to 276 monthly observations over the 23-year period [9].

It was used the time series monthly guest nights registered in hotels for the reason that is one of the measure that help to quantify the national (domestic) and international tourism demand simultaneous.

Monthly Guest Nights in Hotels for both tourism destinations are shown in Fig. 1, so that it can easily be seen from their pattern that there are irregular oscillations suggesting a non-stabilisation of the average and the presence of seasonality, maximum values in the summer months and minimum values in the winter months.

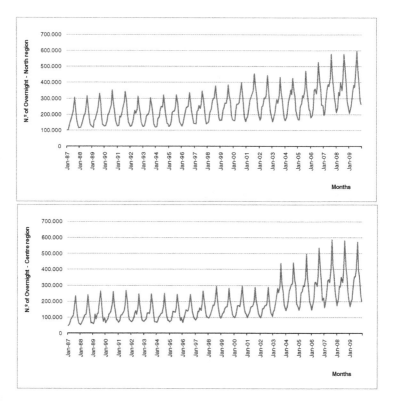

Fig. 1 Monthly Guest Nights in the North and Centre regions of Portugal, from 1987:01 to 2009:12

The Artificial Neural Networks model used in this study was the standard three-layer feedforward network. Since the one-step-ahead forecasting is considered, only one output node is employed. The activation function for hidden nodes is the logistic function; and for the output node the identity linear function. Bias terms are used in both hidden and output layer's nodes. The Resilient Backpropagation algorithm [18] was employed in the training process. The ANN was randomly initialised with weights and bias values. The selection of the architecture, activation function and training algorithm was supported in the previous author's work. For architecture selection several experiments with different architectures were carried out (train and test) and the best architectures were selected according to the results in a validation set using hundreds of training sessions. The selected architecture consists of 12 input nodes in the entrance layer, 4 hidden nodes in the second layer and one node in the output layer (1-12;4;1). The input of the model consists of the 12 previous numbers, corresponding to the last 12 months overnights. The output is the predicted overnights for the following month.

The time series with the original data were divided into three distinct sets: the training set (228 observations for both series, corresponding to the period between January 1988 and December 2006); the validation set (12 observations, corresponding to the year 2007); and the test set (24 observations, corresponding to the years 2008 and 2009). It should be mentioned that this large test set with 24 months (2 years), are quite unusual and arises a more challenging problem because the error usually becomes larger.

Both series were experimented with the usage of more variables into the input models such, the highest value of the series plus the average of the observed data, in the first phase. In the second phase, the drift - difference - of the peaks was also included in the model besides the previous variables mentioned. But no satisfactory results were obtained.

One of the challenge situations of this work is the difficulty to predict these time series with ANN because it was a general increase during the last years. The ANN gives its output based on the situations used during the training process and these situations have lower values.

In order to minimize this problem the time series was converted to the logarithmic domain. Satisfactory results were obtained for the two series, although these results were not significant. Therefore, the new series that served as the basis for the whole study were the original series GRN and GRC.

For each situations described earlier, 250 training sessions were realised, selecting the results from the best training session and choosing the ANN with the best results in the validation set, for each series. This validation set is used for early stop training by a cross validation process, if the root mean squared error (RMSE) does not decrease in a number of 6 successive training iterations. The use of several training sessions is justified because the initial values of the weights are different in each training session, given also different outputs. Therefore, the best performance is selected along the 250 training processes.

3.2 Results and Analysis

To compare the prediction performance of both approaches for the period between January 2008 to December 2009, the following measures of accuracy were calculated: root mean square error (RMSE), correlation coefficient (r) and mean absolute percentage error (MAPE). This last criterion was based on classification proposed by Lewis [12].

Table 1 shows the empirical results for both models and regarding the performance in the test set presented by RMSE and r, we can say that the final results are stable and reached an interesting and satisfactory performance. According to the criterion of MAPE for Model Evaluation in Lewis [12], the predicted data with the selected model has a highly accurate forecast, because the results are lower than 10%. When the MAPE was calculated for the test set, for each of the regions, it was seen that, for the North region, the ANN model presented a value of 7.32%, for 2008 year, and 5.42%, for 2009 year. It is remarkable the quite low MAPE for the second year. Similar values were also produced for the Centre region, 7.01% and 9.33%, for 2008 and 2009 years, respectively. This fact confirms the ability of the model to produce similar performance for different time series.

Table 1 Results of ANN models

Tourism Destination	Forecast Year	Performance Measured - Test Set		
		r	RMSE	MAPE
GRN ANN(1-12;4;1)	**2008**	0.968	30 323	7.32%
	2009	0.983	24 701	5.42%
GRC ANN(1-12;4;1)	**2008**	0.975	26 456	7.01%
	2009	0.960	35 850	9.33%

Fig. 2 shows the predicted and target values in both time series.

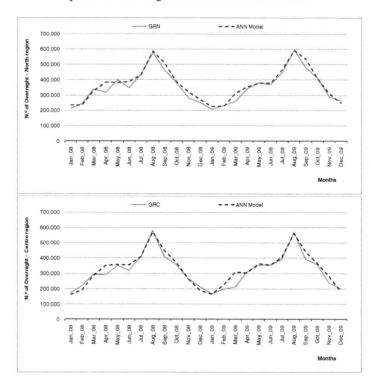

Fig. 2 Comparison between Original Data and Predicted Values with ANN models for both regions, from 2008:01 to 2009:12

However, for both regions there was a significant gap in some months between the forecast values and those that were actually observed, resulting from the fact that the models showed some difficulty in making good forecasts whenever events occurred that caused them to significantly alter the observed values, despite their continuing to be classified as reliable forecasts. These facts may, for example, be a consequence of the high level of promotion in international markets that has been afforded to the regions under analysis. At the same time, local authorities have also invested more heavily in the promotion and organisation of cultural events and holding theme-based trade fairs, amongst other events. For the North region, investments were made in the promotion of some tourist destinations, such as the Douro International Natural Park and the Alto Douro Wine Region, while in the Centre region attention was paid to promoting and investing in the creation of better facilities for winter sports, namely skiing and snowboarding, which attract people to the region, especially in the winter months. Since they were not incorporated into the models, all these factors mean that the models themselves have some difficulty in producing forecasts.

4   Conclusion and Future Developments

This research provides a new and effective ways for quantitative prediction of tourism research, for two tourism destinations - the North and Centre regions of Portugal, using the Artificial Neural Networks (ANN) methodology. These prediction models can be useful for the tourism management and investment decisions.

The forecast values for future tourism demand for the years 2008 and 2009 was presented and analysed, and then compared with the real values.

The models based on the ANN methodology consists on a feedforward structure and trained with the resilient backpropagation, has 4 neurons in the hidden layer and the logistic sigmoidal activation function was used. The input of the model consists of the 12 previous numbers - corresponding to the last 12 months overnights. The output is the predicted overnights for the next month. Monthly Guest Nights in Hotels in both regions of Portugal were used for model training and testing. Experimental results indicated that the ANN models attained high forecasting accuracy. It can be concluded that the ANN models, can be used with satisfactory statistical and adjustment qualities, showing that they are suitable for modelling and forecasting the reference time series. The ANN methodology suggested that this method should be used in time series with non-linear behaviour.

In future, because the models showed some difficulty in making good forecasts for some events/months, it is suggested that it should include other economic variables and dummy variables for this purpose. The authors already have experimented with success the usage of the monthly hours of sun to model the no previously programmed visits [20]. In order to model the general increase of the time series a time index was also used by the author with success [20].

Moreover, the models can be extended to cover more tourism destinations in Portugal and even predict the total expected Guest Nights in Hotels for Portugal or for a particular Hotel.

References

1. Basheer, I., Hajmeer, M.: Artificial Neural Networks: fundamentals, computing, design and application. Journal of Microbiological Methods 153, 3–31 (2000)
2. Bishop, C.M.: Neural Networks for pattern recognition. Oxford University Press, Oxford (1995)
3. Fernandes, P.: Modelling, Prediction and Behaviour Analysis of Tourism Demand in the North of Portugal, Ph.D. Thesis in Applied Economy and Regional Analysis, Valladolid University - Spain (2005)
4. Fernandes, P., Teixeira, J.: A new approach to modelling and forecasting monthly overnights in the Northern Region of Portugal. In: Proceedings of the 4th International Finance Conference. Université de Cergy, Hammamet (2007)
5. Fernandes, P., Teixeira, J.: New Approach of the ANN Methodology for Forecasting Time Series: Use of Time Index. In: Proceeding of ICTDM, Kos-Greece (2009)
6. Fernandes, P., Teixeira, J., Ferreira, J., Azevedo, S.: Modelling Tourism Demand: A Comparative Study between Artificial Neural Networks and the Box-Jenkins Methodology. Journal of Economic Forecasting 5(3), 30–50 (2008)
7. Haykin, S.: Neural Networks. A comprehensive foundation. Prentice-Hall, New Jersey (1999)
8. Hill, T., O'connor, M., Remus, W.: Neural network models for time series forecasts. Management Science 42(7), 1082–1092 (1996)
9. INE, National Institute of Portuguese Statistics, Lisbon (1987-2009)
10. Law, R.: Back-propagation learning in improving the accuracy of neural network-based tourism demand forecasting. Tourism Management 21, 331–340 (2000)
11. Law, R., Au, N.: A neural network model to forecast Japanese demand for travel to Hong Kong. Tourism Management 20, 89–97 (1999)
12. Lewis, C.: Industrial and Business Forecasting Method. Butterworth Scientific, London (1982)
13. Lin, C., Chen, H., Lee, T.: Forecasting Tourism Demand Using Time Series, Artificial Neural Networks and Multivariate Adaptive Regression Splines: Evidence from Taiwan. International Journal of Business Administration 2(2), 14–24 (2011)
14. Machado, T., Teixeira, J., Fernandes, P.: Modelação da Procura Turística em Portugal - Regressão Linear versus Redes Neuronais Artificiais. Revista Turismo & Desenvolvimento 13, 325–335 (2010)
15. Muzaffer, U., Sherif, M.: Artificial Neural Networks versus Multiple Regression in Tourism Demand Analysis. Journal of Travel Research 38(2), 111–118 (1999)
16. Palmer, A., Montano, J., Sese, A.: Designing an artificial neural network for forecasting tourism time series. Tourism Management 27, 781–790 (2006)
17. Patti, C., Snyder, J.: Using a Neural Network to Forecast Visitor Behavior. Annals of Tourism Research 23(1), 151–164 (1996)
18. Riedmiller, M., Braun, H.: A direct adaptive method for faster back-propagation learning: The RPROP algorithm. In: Proceedings of the IEEE International Conference on Neural Networks (1993)

19. Rumelhart, D., McClelland, J.: Parallel Distributed Processing: Explorations in the Microstructure of Cognition. Foundations, vol. 1. The Massachusetts Institute of Technology Press, Cambridge (1986)

20. Teixeira, J., Fernandes, P.: A Insolação como Parâmetro de Entrada em Modelo Baseado em Redes Neuronais para Previsão da Série Temporal do Turismo. CLME, Maputo (2011)

21. Werbos, P.: Beyond regression: New tools for prediction and analysis in the behavioral sciences. Unpublished doctoral dissertation. Harvard University (1974)

22. Witt, S., Witt, C.: Forecasting tourism demand: a review of empirical research. International Journal of Forecasting 11, 447–475 (1995)

23. Zhang, G., Patuwo, E., Hu, M.: Forecasting with artificial neural network: the state of the art. International Journal Forecasting 115, 35–62 (1998)

On Semantic, Rule-Based Reasoning in the Management of Functional Rehabilitation Processes

Laia Subirats, Luigi Ceccaroni, Cristina Gómez-Pérez, Ruth Caballero, Raquel Lopez-Blazquez, and Felip Miralles

Abstract. A clinical decision support system, based on rules described in the semantic web rule language and with semantic annotations from biomedical and time ontologies, is used to reason on processes modeled in the business process modeling notation. This paper, as a case study within the framework of functional rehabilitation processes, analyzes the modeling of the rehabilitation activity consisting of improving the upper limb functioning of patients. The clinical decision support system provides personalization of therapies and is powerful enough to deal with the special characteristics of a rehabilitation scenario, which includes several types of indicators, medical ontologies, and time annotations of different granularities. This paper presents the main lines of a rule-based, ontological framework to translate informal, descriptive methods about functional rehabilitation with an intuitive semantics to the formal representation needed by computational systems. A rule-based reasoning system is used for the representation of processes' semantics and the modeling categories are based on well-accepted rehabilitation notions. We believe that the solution presented for functional rehabilitation can be generalized to other rehabilitation domains such as respiratory, cognitive and cardiac rehabilitation.

Keywords: Ontologies, rule-based reasoning, rehabilitation processes, disabilities with neurological origin.

Laia Subirats · Luigi Ceccaroni · Felip Miralles
Barcelona Digital Technology Centre, 08018 Barcelona, Spain
e-mail: {lsubirats,lceccaroni,fmiralles}@bdigital.org

Cristina Gómez-Pérez · Raquel Lopez-Blazquez
Guttmann Institut Hospital for Neurorehabilitation, 08916 Badalona, Spain
e-mail: {cgomez,rlopez}@guttmann.com

Ruth Caballero
Biomedical Engineering and Telemedicine Group of the Technical University of Madrid,
28040 Madrid, Spain
e-mail: rcaballero@gbt.tfo.upm.es

J. Casillas et al. (Eds.): *Management Intelligent Systems*, AISC 220, pp. 51–58.
DOI: 10.1007/978-3-319-00569-0_7   © Springer International Publishing Switzerland 2013

1 Introduction

In medical rehabilitation there are procedures that are set up with the intent of increasing efficiency, consistency and quality. Unfortunately, the degree to which these procedures are followed often is influenced by various aspects of the procedures, and of the individuals performing the tasks or managing the procedures. Such aspects include procedures' ambiguity, human errors and inconsistency, urgency, patient anxiety. The concepts of process engineering and workflows embody the ideas of controlling and coordinating these complex procedures, activities and interactions among individuals and software components [2].

The objective of the research described in this paper is an improved quality and efficiency of computer-supported work, and specifically the management of rehabilitation processes and the development of a new model for interoperable healthcare, in which data from heterogeneous sources are integrated through workflows that cover the full cycle from diagnosis to treatment. On a lower level, this new model will also facilitate semantic search, discovery, quality assessment, transformation, access, aggregation, analysis of information resources, and publication of derived results for new use. Crucially, the model allows interoperability across scientific domains and user-types (from expert to non-expert) by documenting not only data but also how the data are used for different purposes through workflows, which are then composed as service chains and made reusable. Whilst some progress in this direction has been made in recent research like the one by Fry and Sottara [3], Wieringa et al. [13], and Smith et al. [12], the extension to include administrative, sensor and clinical data is novel, as will be shown in the application to medical rehabilitation. The paper illustrates how to effectively involve healthcare people in the development of a model for rehabilitation processes, to contribute to the evolution from the current situation to individualized rehabilitation and to solve interoperability problems. This is achieved applying rule-based reasoning to decision support and providing a framework that allows the management of semantic enrichment of process models produced by healthcare experts. Related initiatives include: using ontologies to formalize care actions from clinical guidelines [11][7][8]; integrating temporal and resource constraints to generate patient-tailored treatment plans [5]; and using rule-based frameworks to construct complex temporal queries [10]. However, none of these deal with several characteristics of a rehabilitation scenario: (1) multiple types of indicators: session result-indicators, treatment result-indicators and process indicators; (2) multiple granularities of indicator-behaviors in time trends; (3) multiple medical ontologies, such as the International Classification of Functioning, Disability and Health (ICF), the Systematized Nomenclature of Medicine Clinical Terms (SNOMED CT), or the International Classification of Diseases (ICD) version 10 and version 11; (4) semantic annotations from time ontologies.

2 Scenario

As an example, Minerva is the (anonymized) name of a 25-year-old girl who had a car accident while she was biking, some months ago. Now she is at Institut Guttmann, a neurorehabilitation hospital, performing rehabilitation to improve social participation, performance of activities of daily living and, in general, functionality that was impaired due to her injury. Specifically, Minerva suffers from complete spinal cord injury (G95.9 and T09.3 in the ICD-10) at the C6 spinal cord segment, and the objective of her rehabilitation is to improve her upper limb functioning (movement functions (b750-b789) and structure of upper extremity (s730), according to the ICF) executing activities of daily living (ADLs).

In the rehabilitation of upper limb functioning by executing ADLs, several technologies are used: a T-shirt and a robotic orthosis with sensors, which allow therapists to assess the correct execution of an activity and assist patients' movement when needed; a virtual-reality environment, which gives the patient guidance and visual/hearing feedback during the execution of an activity (The orthosis provides the patient sensory stimuli to enable the interaction with the virtual environment.); and a clinical decision support system (CDSS) [4]. Using this CDSS, a personalized therapy for Minerva, including the Bottle-shelf activity (which consists of taking a bottle from a shelf, putting it on a table, and returning it to the start point; see Fig. 1 and below), is recommended. In doing so, the tool takes into account that disease of spinal cord, unspecified (G95.9) is a subclass of diseases of the nervous system (G00-G99 in the ICD-10) and that there are no contraindications for the patient.

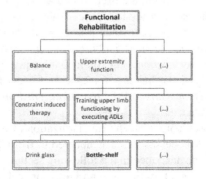

Fig. 1 *Functional-rehabilitation* processes and activities: examples of activities are on the bottom row of the tree shown and processes are the other nodes

3 Methodology

The semantic, rule-based framework we propose uses standard annotations from biomedical and time ontologies and is validated using neurorehabilitation *processes* from clinical practice (see Fig. 2). These processes are composed of

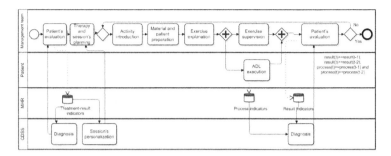

Fig. 2 General workflow of a rehabilitation process involving ADLs using the *business process modeling notation* (BPMN)

rehabilitation *activities*, such as the *Bottle-shelf* activity, which is included in the *Training upper limb functioning by executing ADLs* process, which is a subprocess of *Upper extremity function*.

Activities of the Training upper limb functionality by executing ADLs process, such as Bottle-shelf, are modeled by Caballero et al. [1]. The tasks carried out (by different actors) are: patient's evaluation; therapy and session's planning, activity introduction, material and patient preparation, exercise explanation, ADL execution, exercise supervision, and patient's evaluation. Several indicators are monitored and stored in a medical health record (MHR). *Process indicators*, such as *heart rate*, are used as execution, security or end-session indicators. (A *session* is the time slot in which an activity is executed.) *Session result-indicators*, such as *maximal range of joint movement*, are used as end-criteria of an activity: the therapy continues if, at the end of a session, at least one *session result-indicator* improves with respect to the two previous sessions; otherwise the therapy ends. *Treatment result-indicators*, such as *mobility of joint functions*, *muscle power functions* or *muscle tone functions,* are periodically quantified using measures such as a measure of muscular balance, a measure of upper limb functioning, the *functional independence measure* (FIM), the *spinal cord injury measure* (SCIM) or a measure of joint balance.

The methodology used comprises the standardization of indicators, the computational management of processes initially modeled in BPMN, reasoning and querying, all of which will be described in the following sections.

In the **standardization of indicators and interoperability**, all terminology and annotations used are translated to international standards. The ICF (which standardize attributes and values; values ranging from 0 -no deficiency- to 4 -complete deficiency-), SNOMED CT, ICD-10 and ICD-11 are used for *biomedical annotations* of processes and results. The definition of interoperable indicators is done using the following steps:

1. *Standardization* of treatment result-indicators found on healthcare questionnaires into ICF (together with health professionals).

2. *Inference* of ICF core-set categories corresponding to the indicators. ICF *core sets* are subsets of ICF formed according to functioning, pathology or rehabilitation process. *Core sets* are useful because, in daily practice, clinicians and other professionals need only a fraction of the categories found in the ICF.

3. *Reduction* of the indicators (considering core sets) to no more than 10 ICF categories for each user of the CDSS's interface. The number 10 was set due to human limits in the ability of processing information [9].

4. *Standardization* of *time annotations* using an ontology [10]. This ontology provides *Web ontology language* (OWL) entities for representing propositions, valid times (both instants and intervals), granularity, and duration.

In the **computational management of processes initially modeled in BPMN**, rules are used to define activities and therapies, and to evaluate patients. Initially modeled in BPMN 2.0, they are then coded using the semantic Web rule language (SWRL) [6].In the following examples, a representative set of rules is described in detail[1].) As a first example, let us consider the rule which stops a session if the value of a parameter is too high: *NextTask (?nt), ProcessIndicator (?pi), Patient (?p), hasIndicator (?p, ?pi), greaterThanOrEqual (2, ?pi) → hasNextTask(?p, ?nt)*, which means that, if a patient p has a process indicator pi which is greater than or equal to moderate deficiency (2 in the ICF), the session is stopped and the next task is nt. Similar rules are applied to several process indicators of rehabilitation activities. Process indicators are used, among other things, to stop sessions because of alterations in body functions (such as a too-high heart rate) or environmental factors (such as a too-high temperature or humidity). Rules' conflicts, for example about the conditions to terminate a session, are solved by giving priorities based on: indicators' weight according to ICF core sets, time, and relevance of result indicators with respect to the activity.

Finally, the framework was implemented **reasoning and querying with Pellet and SPARQL**. Reasoning is used for the personalization of rehabilitation therapies, to find the most suitable activities for a patient. The following example shows how the Bottle-shelf activity is selected for a therapeutic plan. The Pellet reasoner is used to infer properties and relationships, while SPARQL (an RDF query language) is used for querying RDF ontologies.

Personalization of therapies. The following set of queries is used to personalize therapies. Firstly, the system evaluates which indicators the patient should improve through the query (1) *Select (?i) where Patient (?p), Indicator (?i), Deficiency (?mild), hasDeficiency (?i, ?mild), has Indicator (?p, ?i)*.

Afterwards, activities which cover this objective are searched for. In the case of Minerva the objective is *power muscles of one limb*, so the query is (2) *Select (?a) where Activity (?a), Power muscles of one limb (?i), hasObjective (?a, ?i)*.

The Bottle-shelf activity is among the results. Then, other activities which can be indicated for the patient based on subclasses of neurological diseases in the

---

[1] All rules are available at http://code.google.com/p/functionalrehabilitation/downloads/list

ICD-10 taxonomy are inferred through Pellet, with the query (3) *Select (?a) where Activity (?a), Patient (?p), has disease (?p, ?d), SubClassOf (?NeurologicalDiseases, ?d).*

Finally, it is checked if the activity is contraindicated to the patient. In the *Bottle-shelf* activity, with the query (4) *Count (?contraindication) where Patient (?p), Moderate, severe or complete neuromusculoskeletal and movement-related functions (?mscn), hasContraindication (?p, ?mscn).*

To be recommended, an activity has to appear as result in queries 2 or 3 and the result in query 4 has to be 0.

Diagnosis. Diagnosis in order to evaluate a patient is performed using temporal reasoning. The system has to deal with two issues: different time annotations for indicators and different kinds of indicators. In order to obtain indicators of the *Upper extremity function* process the following query can be used: *Select (?p, ?resulttreatment, ?week) where Patient (?p), Upper extremity function (?process), hasProcess (?process, ?activity), hasActivity (?p, ?activity), hasIndicator (?p, ?resulttreatment), temporal:overlaps (?activity, ?interval), temporal:after(?interval,"2012-11"), temporal:before(?interval,"2013-3").*

Trend granularities are used in temporal queries in order to measure the intensity of *change*, which has the following domain of values: <increase (I), decrease (D), remain stationary (S)>. Three different levels of trend granularities have been defined in the ontology for the intensity of change: fine, medium and coarse. The parameters used when retrieving indicators are then: *temporal:ValidInterval, temporal:ValidPeriod, temporal:trendGranularity* and *IndicatorType*.

The architecture of the proposed rule-based framework is composed of: (1) a Protégé-based knowledge editor, which uses Jena and SPARQL to interact with the OWL ontology; (2) rules implemented in SWRL; and (3) the Pellet reasoner.

4   Results

Results are based on the application of the proposed framework to the previous scenario. Traceability evaluation is carried out in order to verify the completeness of the rule set. Priorities are given to each indicator and measure to gauge how much the objectives of an activity are achieved.

Let us suppose that Minerva is performing the Bottle-shelf activity and her heart rate is too high. As a consequence, an alarm is triggered and her session is stopped. After the alarm, her therapist wants to evaluate Minerva. She wants to diagnose her in the upper extremity function process with a 3-weeks granularity. From the tool, see Fig. 3, a summary of indicators of body functions, body structures, activities and participation, and environmental factors are summarized; deficiency and difficulty levels are represented as red/4 (complete), orange/3 (severe), yellow/2 (moderate), green/1 (mild) and blue/0 (no).

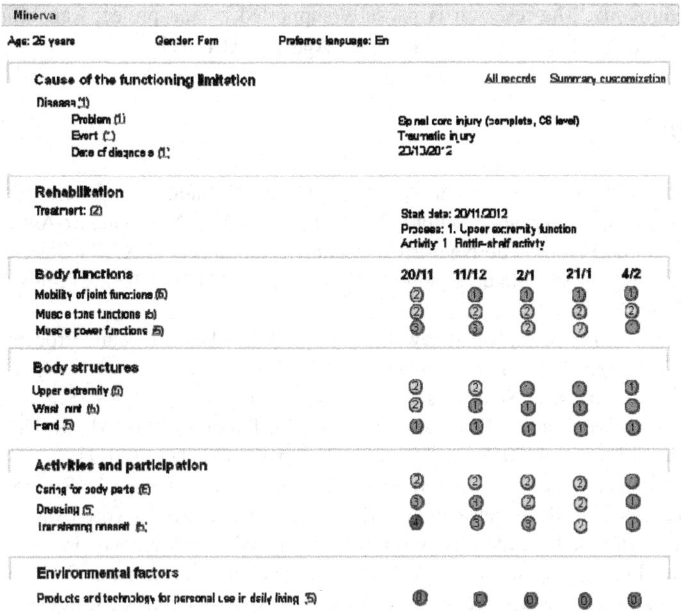

Fig. 3 CDSS's interface of Minerva's therapy and diagnosis

5 Conclusions

Temporal and biomedical annotations can provide semantics to functional rehabilitation processes represented in BPMN 2.0. This semantics, together with rules obtained from literature and data mining, can be used to introduce automatic reasoning in decision support. We present a semantic, rule-based reasoning framework which uses existing temporal ontologies and formal processes notation to enhance interoperability and reasoning in rehabilitation: extending temporal ontologies, specifying granularities of indicator-behavior in time, working with different types of indicators and merging medical ontologies such as ICF, SNOMED CT, ICD-10 and ICD-11.

Different issues appear when there are large amount of rules in BPMN 2.0 processes which are solved generalizing rules; while problems of conflicting rules are solved by establishing priorities. Furthermore, time-trend granularities are added to temporal ontologies to construct temporal queries in order to measure the intensity of change.

The proposed CDSS is validated with a functional rehabilitation scenario and the correct execution of functional rehabilitation processes. Finally, the graphical user interface is validated by clinicians' opinions. We believe that the solution presented for functional rehabilitation, has implications in an improved quality and efficiency of management of functional rehabilitation processes, and that can be generalized to other rehabilitation domains such as respiratory, cognitive and cardiac rehabilitation.

Acknowledgments. The research is partially supported by the project Rehabilita (CEN-2009-1043) and the Catalonia Competitiveness Agency (ACC1Ó).

References

1. Caballero-Hernández, R., Gómez-Pérez, C., Cáceres-Taladriz, C., García-Rudolph, A., Vidal-Samsó, J., Bernabeu-Guitart, M., Tormos-Muñoz, J.M., Gómez-Aguilera, E.J.: Modelado de Procesos de Neurorrehabilitación. In: Actas del XXIX Congreso Anual de la Sociedad Española de Ingeniería Biomédica (CASEIB 2011), Cáceres (España),, pp. 125–128 (2011)
2. Cichocki, A., Helal, A., Rusinkiewicz, M., Woelk, D.: Workflow and Process Automation. Kluwer Academic Publishers (1998)
3. Fry, E., Sottara, D.: Standards, Data Models, Ontologies, Rules: Prerequisites for Comprehensive Clinical Practice Guidelines. In: Palmirani, M. (ed.) RuleML 2011 - America. LNCS, vol. 7018, pp. 252–266. Springer, Heidelberg (2011)
4. Gómez-Pérez, C., Caballero-Hernández, R., Medina-Casanovas, J., Roig-Rovira, T., Vidal-Samsó, J., Bernabeu-Guitart, M., Cáceres-Taladriz, C., Tormos-Muñoz, J.M., Gómez-Aguilera, E.J.: Identificación de Oportunidades de Mejora en Procesos de Neurorrehabilitación. In: Actas de CASEIB 2012, San Sebastián, España (2012)
5. González-Ferrer, A., ten Teije, A., Fdez-Olivares, J., Milian, K.: Careflow planning: From time-annotated clinical guidelines to temporal hierarchical task networks. In: Peleg, M., Lavrač, N., Combi, C. (eds.) AIME 2011. LNCS, vol. 6747, pp. 265–275. Springer, Heidelberg (2011)
6. Horrocks, I., Patel-Schneider, P.F., Boley, H., Tabet, S., Grosof, B., Dean, M.: Swrl: A semantic web rule language combining owl and ruleml. W3c member submission (2004)
7. Jafarpour, B., Abidi, S.R., Abidi, S.S.R.: Exploiting OWL reasoning services to execute ontologically-modeled clinical practice guidelines. In: Peleg, M., Lavrač, N., Combi, C. (eds.) AIME 2011. LNCS, vol. 6747, pp. 307–311. Springer, Heidelberg (2011)
8. Kashyap, V., Morales, A., Hongsermeier, T.: On implementing clinical decision support: achieving scalability and maintainability by combining business rules and ontologies. In: AMIA Annu. Symp. Proc., pp. 414–418 (2006)
9. Miller, G.A.: The magical number seven, plus or minus two: Some limits on our capacity for processing information. Psychological Review 63(2), 8197 (1956)
10. O'Connor, M.J., Hernandez, G., Das, A.: A rule-based method for specifying and querying temporal abstractions. In: Peleg, M., Lavrač, N., Combi, C. (eds.) AIME 2011. LNCS, vol. 6747, pp. 255–259. Springer, Heidelberg (2011)
11. Pruski, C., Bonacin, R., Da Silveira, M.: Towards the formalization of guidelines care actions using patterns and semantic web technologies. In: Peleg, M., Lavrač, N., Combi, C. (eds.) AIME 2011. LNCS, vol. 6747, pp. 302–306. Springer, Heidelberg (2011)
12. Smith, F., Missikoff, M., Proietti, M.: Ontology-based querying of composite services. In: Ardagna, C.A., Damiani, E., Maciaszek, L.A., Missikoff, M., Parkin, M. (eds.) BSME 2010. LNCS, vol. 7350, pp. 159–180. Springer, Heidelberg (2012)
13. Wieringa, W., op den Akker, H., Jones, V.M., op den Akker, R., Hermens, H.J.: Ontology-based generation of dynamic feedback on physical activity. In: Peleg, M., Lavrač, N., Combi, C. (eds.) AIME 2011. LNCS, vol. 6747, pp. 55–59. Springer, Heidelberg (2011)

Advances in Market Segmentation through Nature-Inspired Intelligence Methods: An Empirical Evaluation

Charalampos Saridakis, Stelios Tsafarakis, George Baltas, and Nikolaos Matsatsinis

Abstract. Market segmentation is a broadly recognized concept in strategic marketing and planning. Although K-means cluster analysis has been traditionally used as a means to segment markets during the last 50 years, the results have often been reported to be less than satisfactory. This paper develops and introduces a new nature-inspired mechanism, called Particle Swarm Optimization, to the problem of market segmentation. The proposed mechanism, which addresses shortcomings of existing approaches, is being implemented in an empirical dataset of 1,622 consumers pertaining to attribute-level satisfaction ratings from currently owned cars. The results are encouraging and provide decision makers with improved alternatives over existing market segmentation methods.

Keywords: Market segmentation, Cluster analysis, Particle Swarm Optimization, K-means clustering.

1   Introduction

Market segmentation is a fundamental concept of modern marketing and is one of the most pervasive activities in both the marketing academic literature and practice [2]. This key strategic concept is widely used as a means to distinguish

Charalampos Saridakis
Leeds University Business School, University of Leeds, United Kingdom
e-mail: b.saridakis@leeds.ac.uk

Stelios Tsafarakis · Nikolaos Matsatsinis
Department of Production Engineering & Management,
Technical University of Crete, Greece
e-mail: tsafarakis@isc.tuc.gr, nikos@ergasya.tuc.gr

George Baltas
Department of Marketing & Communication, Athens University of Economics & Business,
Greece
e-mail: gb@aueb.gr

J. Casillas et al. (Eds.): *Management Intelligent Systems*, AISC 220, pp. 59–66.
DOI: 10.1007/978-3-319-00569-0_8   © Springer International Publishing Switzerland 2013

homogeneous groups of customers who can be targeted in the same manner because they have similar needs and preferences [10]. A good segmentation solution will result in segment members that are as similar as possible within the segment, and as dissimilar as possible between the segments. Marketers generally agree that if segmentation is properly applied, it would guide companies in tailoring their product and service offerings to the groups most likely to purchase them.

Despite the appealing benefits of market segmentation, recent results [12] show that there are implementation problems that lead to its disappointing performance in many real-world applications. More importantly, the progress in clustering algorithms is too limited over the last decades. Even though K-means was first proposed over 50 years ago, it still remains one of the most widely used clustering algorithms for market segmentation due to its simplicity, ease of implementation, speed, efficiency, and empirical success. The main limitation of K-means is that the final result depends on the selection of the initial cluster centers (centroids). Different initializations lead to different final clustering partitions because the algorithm searches the local optimal solution in the vicinity of the initial solution. The same initial centroids will always produce the same cluster results.

Against this background, the aim of this paper is twofold. First, to overcome limitations of existing mechanisms, by introducing a new Particle Swarm Optimization (PSO) algorithm for clustering market data. Instead of just refining the position of the centroids at each iteration, the PSO clustering algorithm performs a globalized searching, starting from a wide range of different initial cluster centroids. Second, to apply and evaluate this method in an empirical dataset collected through a large-scale survey research examining consumer buying behaviour in the car market. An attempt has been made to present and empirically evaluate a new tool for clustering and market segmentation. The importance of our study from a managerial perspective is evident, since good market segmentation mechanisms contribute to a full understanding of the market, the ability to predict behaviour with great precision, an increased likelihood of detecting and exploiting new market opportunities and the identification of the groups worth pursuing [8]. Our proposed segmentation mechanism is used to segment a total of 1,622 respondents and identify clusters of homogenous consumer groups regarding their level of satisfaction from their current car's attribute-level performance.

2 K-Means Clustering

The K-means algorithm works as follows (Jain, 2010). Let $X=\{x_i\}$, $i=1,\ldots,n$ be the set of d-dimensional points to be clustered into a set of K clusters $C=\{c_k\}$, $k=1,..,K$. The algorithm finds a partition such that the squared error between the empirical mean of a cluster c_k (centroid μ_k) and the points that belong to the cluster is minimized. The goal of K-means is to minimize the sum of the squared error over all K clusters:

$$J(C) = \sum_{k=1}^{K} \sum_{x_i \in c_k} \left\| x_i - \mu_k \right\|^2 \tag{1}$$

The main steps of K-means algorithm are:

1. Select (usually at random) K initial cluster centroids.
2. Form the initial clusters by assigning each data point to its closest centroid.
   repeat
3. Compute the new cluster centroids as the mean value of all the data points that belong to each cluster.
4. Generate the new clusters by assigning each data point to its closest centroid.
   until cluster membership stabilizes.

3 The Particle Swarm Optimization Algorithm

Particle Swarm Optimization is a population-based swarm intelligence algorithm. It was originally proposed by Kennedy and Eberhart [3] as a simulation of the social behavior of social organisms, such as bird flocking and fish schooling.

The PSO algorithm works as follows. First, a set of P particles (population) is randomly initialized, where the position of each particle corresponds to a solution of the problem, represented by a d-dimensional vector in the problem space $s_i = (s_{i1}, s_{i2},..., s_{id})$, $i = 1, 2,..., P$, $s \in \Re$. Thus each particle is randomly placed in the d-dimensional space as a candidate solution, and its performance is evaluated on the predefined fitness function. The velocity of the i-th particle $v_i = (v_{i1}, v_{i2} ,..., v_{id})$, $v \in \Re$, is defined as the change of its position in each algorithm's iteration. The algorithm completes the optimization through following the personal best solution of each particle and the global best value of the whole swarm. Each particle adjusts its trajectory toward its own previous best position and the previous best position attained by any particle of the swarm, namely p_{id} and p_{gd}. The velocities and positions of particles are updated using the following formulas:

$$v_{id}(t+1) = v_{id}(t) + f_1 rand_1(p_{id} - s_{id}(t)) + f_2 rand_2(p_{gd} - s_{id}(t)) \tag{2}$$

$$s_{id}(t+1) = s_{id}(t) + v_{id}(t+1) \tag{3}$$

where t is the iteration counter, *f1* and *f2* are the acceleration coefficients, *rand1*, *rand2* are two random numbers in [0, 1]. The acceleration coefficients *f1* and *f2* control how far a particle will move in a single iteration, and are typically both set to 2. The newly formed particles are evaluated according to the objective function and the algorithm iterates for a predetermined number of generations (iterations), or until a convergence criterion has been met. Finally, the best solution obtained across all generations is returned.

4 PSO Clustering

Several works of PSO applications to data clustering have been reported in the literature (e.g., [9][1][4]). Specifically, the PSO algorithm is used to cluster well-known databases (e.g. Iris), or artificial data sets, and its performance is compared to other clustering algorithms (usually K-means). Based on these works, we implement a PSO clustering algorithm in Matlab, and apply it to market segmentation using a real world data set. In our implementation each particle represents the centroids of the K clusters of the problem. Hence each particle has a length of $l=K*d$ where d is the number of dimensions in the problem space. For instance, in a problem with two clusters, where the number of dimensions in each data point is three, the particle |1.4 -0.7 -1.1 0.2 2.3 -0.9| represents a possible solution to the problem where the first cluster centroid is (1.4 -0.7 -1.1) and the second is (0.2 2.3 -0.9). Each particle s_i represents a candidate solution, and the whole swarm represents a group of solutions that collectively search for the global optimum.

Following van der Merwe and Engelbrecht (2003), we use the minimization of the quantization error as the problem's objective function, thus the fitness of each particle is calculated with the following formula:

$$J_e = \frac{\sum_{k=1}^{K}\left[\sum_{x_i \in c_k}\|x_i - m_k\|^2 / N_k\right]}{K}$$

where m_k is the centroid of the k-th cluster c_k, and N_k is the number of data points that belong to cluster c_k. The PSO clustering algorithm works as follows:

Initialization
1. Generate the initial population where each particle contains K randomly generated centroids
2. **For** each particle (*Evaluation of the particles' fitness*)
3.   **For** each data vector (*Form the clusters*)
4.   Calculate the Euclidean distance to all cluster centroids
5.   Assign the data vector to the closest centroid
6.   **EndFor**
7.   Calculate the particle's fitness according to the objective function 4
8. **EndFor**
9. **Keep** the optimum solution of each particle
10. **Keep** the optimum particle of the whole swarm
Main Phase
11. **Do until** the maximum number of iterations has been reached
12.   Calculate the velocity of each particle according to function 2
13.   Calculate the new position of each particle according to function 3
14.   Evaluate the new fitness of each particle (steps 2. to 8.)
15.   Update the optimum solution of each particle and the optimum particle of the whole swarm
16. **Enddo**
17. **Return** the best solution/set of solutions.

5   Comparison of the Two Algorithms

In order to assess the proposed PSO clustering algorithm we compare its perfor-mance with that of K-means using data from a large-scale survey research examin-ing consumer preferences and buying behavior in the car market. A total of 1,622 respondents from the population of a European metropolitan area participated in the survey. Among others, individual data on consumer's current car's attribute-level performance was collected for a broad range of 36 car characteristics. Attribute-level performance was measured on a 5-point scale, ranging from *not at all satisfied* (1) to *extremely satisfied* (5). Although multicollinearity might not be a serious issue for a study considering a limited number of attributes, when model-ing a large number of attributes, it could become difficult to detect and control for (e.g., [5][11]). In this direction, factor analysis and varimax rotation were used to reduce the number of attributes and to identify composite attributes based on res-pondents' evaluations. In total, seven factors with eigenvalues greater than one were extracted from the 36-item instrument. With these seven factors, 60% of the total variance was explained, with construct reliabilities (i.e., cronbach's alpha values) of 0.90, 0.91, 0.83, 0.86, 0.73, 0.72, and 0.70. The factor of "quality and safety" includes items such as assembly quality, quality of materials, technology, reliability and road behaviour. The factor of "performance and technical character-istics" includes items such as engine capacity, horsepower, maximum speed, and acceleration. The factor of "spaciousness" includes items such as overall roomi-ness, luggage roominess and cabin roominess. The factor of "intangible and image characteristics" includes items such as country of origin, manufacturer image, and overall style and design. The factor of "secondary cost related characteristics" in-cludes items such as taxes and dues, insurance premium, and fuel consumption. The factor of "manoeuvrability characteristics" includes items such as dimensions, weight, and manoeuvrability in the city. Finally, the factor of "primary cost related characteristics" includes items such as purchase price, services and maintenance cost, and warranties. The seven factors were used as clustering variables to assess the performance of the proposed PSO clustering algorithm to that of K-means.

The length of each particle is $7*K$, depending on the number of clusters (K) that are specified for the problem. The number of particles (population size) is 30, the number of algorithm's iterations is 50, and *f1* and *f2* are both set to 2. These values came after testing a wide range of different values in the application of the PSO algorithm to the data set using different number of clusters. Since both algorithms require the number of clusters to be specified by the user, we first apply a hierar-chical clustering approach following Punj & Stewart [7]. We use the Euclidean distance as a function of measuring the similarity between every pair of objects in our data set, and the Ward algorithm (inner squared distance) for grouping the ob-jects into clusters. The best solution reached by the application of the hierarchical clustering resulted was for $K= 3$ clusters, which constitutes the number of different market segments. The performance of the two algorithms is evaluated with the use of the "silhouette" method. The silhouette value for each data point is a measure of how similar that point is to the other points in its own cluster versus points in other clusters. Its value ranges from -1 (worst) to +1 (best). The silhouette value of data point i is defined as

$$Sh(i) = \frac{\min[aod(i,k)] - aid(i)}{\max\{aid(i), \min[aod(i,k)]\}}$$

where $aid(i)$ is the average distance from the i-th point to the other points in its own cluster (Average Inner Distance), and $aod(i,k)$ is the average distance from the i-th point to points in another cluster c_k (Average Outer Distance). The cluster c_k with the lowest aod from point i ($\min[aod(i,k)]$) is called the "neighbouring cluster". After calculating the silhouette values i for all the points, we estimate the mean value of Sh in the data set, which constitutes a measure of how appropriately the data has been clustered. Values of mean(Sh) close to 1 indicate that all the data in the cluster are tightly grouped. We also calculate the Sum of Euclidean Distances (Kuo et al., 2011):

$$SED = \sum_{i=1}^{n} \sum_{j=1, x_i \in cj}^{K} d(x_i, m_j), \text{ where } d \text{ the Euclidean distance.}$$

We run the two algorithms for 20 replications. In each replication the K-means algorithm is executed 10 times and only the best solution is maintained, in an attempt to mitigate the dependence of the final result from the initial centroids. The PSO algorithm is executed once in each replication. The average value for the mean(Sh) across the 20 replications is 0.2032 for the K-means and 0.2128 for the PSO. The respective values of SED (the smaller the SED the better the result) are 11972 and 11337. The nature inspired approach performs slightly better, despite the fact that K-means is executed 10 times in each replication. K-means converges in 1-2 seconds, whereas PSO needs about 13-15 seconds. Table 1 reports the clustering scheme along with the respective cluster centroid values for the best solution provided by the PSO throughout the 20 replications. The same clustering scheme is also graphically depicted in Figure 1.

Table 1 Clustering scheme and centroid values

	No. data	Cluster Centroids						
		Quality & safety	Performance & technical	Spaciousness	Intangible & image	Secondary cost related	Maneuverability	Primary cost related
Cluster 1	531	-0.322	-0.722	-0.212	-0.098	-0.007	-0.794	0.086
Cluster 2	758	0.205	0.396	0.616	0.048	0.275	0.214	-0.093
Cluster 3	333	0.045	0.249	-1.064	0.046	-0.616	0.780	0.076

A look at the cluster solution suggests that results are highly interpretable. Cluster 1 consists of 531 respondents and includes all those consumers who are generally dissatisfied with the performance of most of their current car's characteristics, although they seem to be less dissatisfied with their car's cost related characteristics (primary and secondary). In contrast, cluster 2, which consists of 758 respondents, includes those consumers who are relatively satisfied with the performance of most of their current car's characteristics (and at the same time less satisfied with their car's primary cost related characteristics). Finally, cluster 3 consists of 333 respondents and includes those consumers who seem to be extremely satisfied with some aspects of their cars (e.g., maneuverability) and extremely dissatisfied with same others (e.g., spaciousness). It could be suggested that cluster 1 represents owners of low-end car models who are generally dissatisfied with most characteristics of their current car, cluster 2 represents owners of high-end car models who are generally satisfied with most characteristics of their current car, and cluster 3 represents more critical consumers who probably possess mainstream, mass-market car models.

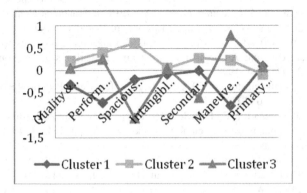

Fig. 1 Clustering scheme

6 Conclusions

Important limitations of existing widely used segmentation algorithms require the introduction of new segmentation methods that will address computational shortcomings. Against this background, this paper illustrated a new approach, namely Particle Swarm Optimization, for the market segmentation problem. Our mechanism is inspired from the collective behavior of social organisms found in nature, such as bird flocking and fish schooling. The preceding analysis employs a biological metaphor derived from collective systems and determines market-driven segments that are evolved in such a way that segment members are as similar as possible within the segment, and as dissimilar as possible between the segments. The search in this collective system of swarm intelligence is driven by cooperation among candidate solutions. The approach was illustrated through an application to an empirical dataset of 1,622 consumers pertaining to attribute-level satisfaction ratings from currently owned cars. Although more research on this innovative segmentation mechanism and additional empirical evaluation in different business settings are further topics to be investigated, the paper provides clear insights into how PSO can be applied to identify optimal market segments.

The approach presented here has important implications for marketing researchers and practitioners who may want to revise their current practices for market segmentation. We acknowledge the fact that the application of the proposed new state-of-the-art, nature-inspired clustering method might cause a great deal of work, but it is likely that improved results, accurate exploitation of new market opportunities and precise identification of the groups worth pursuing should easily outweigh these costs. It is hoped that the ideas presented here are demonstrative of the potential of artificial intelligence-based solutions to such decisional situations.

References

1. Chen, C.Y., Ye, F.: Particle swarm optimization algorithm and its application to clustering analysis. In: Proceedings of the 2004 IEEE International Conference on Networking, Sensing and Control, Taipei, Taiwan, pp. 789–794 (2004)
2. DeSarbo, W., Grisaffe, D.: Combinatorial optimization approaches to constrained market segmentation: An application to industrial market segmentation. Marketing Letters 9(2), 115–134 (1998)
3. Jain, A.K.: Data clustering: 50 years beyond K-means. Pattern Recognition Letters 31(8), 651–666 (2010)
4. Kennedy, J., Eberhart, R.C.: Particle Swarm Optimization. In: Proceedings of the IEEE International Conference on Neural Networks, vol. IV, pp. 1942–1948. IEEE Service Center, Piscataway (1995)
5. Kuo, R.J., Wang, M.J., Huang, T.W.: An application of particle swarm optimization algorithm to clustering analysis. Soft Computing 15(3), 533–542 (2011)
6. Mittal, V., Ross, W.T., Baldasare, P.M.: The asymmetric impact of negative and positive attribute-level performance on overall satisfaction and repurchase intentions. Journal of Marketing 62(1), 33–47 (1998)
7. Poli, R., Kennedy, J., Blackwell, T.: Particle swarm optimization. An overview. Swarm Intelligence 1, 33–57 (2007)
8. Punj, G., Stewart, D.W.: Cluster analysis in marketing research: Review and suggestions for application. Journal of Marketing Research 20(2), 134–148 (1983)
9. Tuma, M., Decker, R., Scholz, S.: A survey of the challenges and pitfalls of cluster analysis application in market segmentation. International Journal of Market Research 53(3), 391–414 (2011)
10. van der Merwe, D.W., Engelbrecht, A.P.: Data clustering using particle swarm optimization. In: Proceedings of the 2003 IEEE Congress on Evolutionary Computation, pp. 215–220. IEEE Service Center, Piscataway (2003)
11. Wedel, M., Kamakura, W.: Introduction to the special issue on market segmentation. International Journal of Research in Marketing 19, 181–183 (2002)
12. Wittink, D., Bayer, L.: The measurement imperative. Marketing Research 6, 14–23 (1994)
13. Yankelovic, D., Meer, D.: Rediscovering market segmentation. Harvard Business Review 84(2), 122–132 (2006)

Improving Index Selection Accuracy for Star Join Queries Processing: An Association Rules Based Approach

Benameur Ziani, Ahmed Benmlouka, and Youcef Ouinten

Abstract. Nowadays, the new technologies for Business Intelligence as Data Warehouse, OLAP, Data Mining, emerged and are needed for the managerial process. In the area of decision support systems, a basic role is held by a data warehouse which is an online repository for decision support applications using complex star join queries. Answering such queries efficiently is often difficult due to the complex nature of both the data and the queries. One of the most challenging tasks for the data warhouse administrator (DWA) is the selection of a set of indexes to attain optimal performance for a given workload under storage constraint. The problem is shown to be NP-hard since it involves searching a vast space of possible configurations. It is very much important to extract meaningful information from the workload which represents the major step towards building relevant indexes. This paper presents an approach for selecting an optimized index configuration using association rules with Apriori algorithm which can drive to understand with more accuracy the attributes correlation. This helps to recommend an index set that closely match the requirements of the provided workload. Experimented using the ABP-1 benchmark, our proposed approach achieves good performance compared with previous studies.

Keywords: Business intelligence, Data warehouse, Bitmap join index, Data mining, Association rules.

1 Introduction

The objective of business intelligence is to improve the timeliness and quality of inputs to the decision process. The data warehouse is the significant component of business intelligence [1]. It is a repository of subjectively selected and adapted operational data, which can provide means for implementing effective decision support

Benameur Ziani · Ahmed Benmlouka · Youcef Ouinten
LIM, Computer Science Department, UATL Algeria
e-mail: {bziani,a.benmlouka,ouinteny}@mail.lagh-univ.dz

J. Casillas et al. (Eds.): *Management Intelligent Systems*, AISC 220, pp. 67–74.
DOI: 10.1007/978-3-319-00569-0_9   © Springer International Publishing Switzerland 2013

system. Information of a data warehouse are usually stored using a star schema that is composed of a single fact table and many descriptive dimension tables linked to the fact table through foreign key relationships [2][3]. Although there are several ways to optimize data warehouse systems, implementing an optimized indexing solution is the most effective [4]. The selection of optimal bitmap indexes has always been one of the most important roles of data warehouse administrators (DWAs). Given a workload and a constraint defining the disk space devoted to indexes, the goal is to determine an index set that minimizes the workload execution cost respecting the space constraint. A naive solution, would consider every combination of all potential indexes and then evaluate each resulted configuration. Such a solution is clearly exponential and thus impractical. Hence, most approaches that have been proposed for the bitmap indexes selection problem are mainly focused on pruning its search space by eliminating non relevant attributes. Mostly, the recomendation of an index configuration relies on greedy search algorithms [6]. Obviously, an efficient enumeration of potential useful indexes depends on the attributes utility estimation. Attributes seldom accessed together by queries are likely to be uninteresting because they may not be profitable to the workload. As such, the efficient enumeration of potential indexes directly leads to significant performance improvement.

It is widely recognized that association rules are intuitive, and constitute a powerful tool to identify the regularities and correlation in a set of observed objects. In this paper, we consider the index selection process as a typical association rules mining problem. The indexes are built with combinations of attributes, viewed as items. The queries in the workload, viewed as transactions, are described by the attributes they involve. We aim at representing the correlation between attributes more compactly than by using a greedy search that consider a much larger number of combinations. This reduction of the search space can achieve orders-of-magnitude improvement in the computational complexity of the selection process. In the rest of the paper, we first present background of the studied problem in Section 2. Section 3 provides a survey of related prior work. in Section 4, we outline the proposed approach for the index selection problem. Results of our preliminary evaluation are presented in Section 5. Finally we conclude with some directions for future work in Section 6.

2   Background

This section provides a brief description of the concepts of bitmap join index and association rules mining as a background to our approach.

Bitmap join indexes are efficient data structures for processing complex queries in data warehouse applications. They are proposed to speed up the joins between dimension table(s) and the fact table of relational data warehouses modelled using a star schema [5]. A bitmap join index may be defined on one or several columns, called indexable attributes, of the same dimension table or on more than one dimension table. Given an SQL query, an indexable attribute A for a dimension table D is a non key column D.A such that there is a condition of the form

(D.A θ Expression) in the WHERE clause. The operator θ must be among $(=,<,>,<=,>=,\text{BETWEEN},\text{IN})$. In practice, the number of potential indexes is highly dependent on the number of indexable attributes involved in the query workload. It is not hard to show that the number of possibilities for selecting one bitmap join index is $n_1 = \sum_{i=1}^{k} \binom{k}{i} = 2^k - 1$, where k is the number of indexable attributes. To select more than one bitmap join index, the number of possibilities is $n_2 = \sum_{i=1}^{2^k-1} \binom{2^k-1}{i} = 2^{2^k-1} - 1$. For $k = 7$, we have approximately $n_1 = 127$ and $n_2 = 10^{38}!$.

Given a workload W of n queries $\{q_1^{f_1}, q_2^{f_2}, ..., q_n^{f_n}\}$, where each query $q_i (1 \leq i \leq n)$ has an access frequency f_i, and a storage constraint S, the Bitmap Join Index Selection Problem involves selecting an index configuration C among all possible configurations so that the cost for processing the workload W using C is minimum subject to the limit, on the total indexes size, S.

Obviously, an efficient enumeration of potential useful indexes depends on the attributes utility estimation. Thus, we consider the index selection process as a typical association rules mining problem. Discovering association rules is a key problem in important data mining applications. It aims to extract interesting correlations, frequent patterns or associations among sets of items in the transaction databases. It was originally proposed in [7]. The problem of mining association rules can be formally stated as follows. Let $I = \{i_1, i_2, \ldots, i_m\}$ be a set of m distinct attributes called items. Let D_T be a set of transactions where each transaction T is a set such that $T \subseteq I$. Let X be a set of items. A transaction T is said to contain X if and only if $X \subseteq T$. An association rule is an implication in the form of $X \Rightarrow Y$, where $X \subset I$, $Y \subset I$ and $X \cap Y = \emptyset$. X is called antecedent while Y is called consequent of the rule. There are two important basic measures for association rules, the support s and the confidence c. The rule $X \Rightarrow Y$ has *support* s if $s\%$ of transactions in the transaction set D_T contain $X \cup Y$. The rule $X \Rightarrow Y$ holds in the transaction set D_T with *confidence* c if $c\%$ of transactions in D_T that contain X also contain Y. Since the users concern about only the frequently appeared items, thresholds of support and confidence are predefined to drop those rules that are not so interesting or useful. One of the well-known algorithms for mining association rules is Apriori [7] which utilizes the *support* and *confidence* measures to discover the frequent itemsets satisfying the minimum support value and select strong rules satisfying the minimum confidence value. The input data for mining association rules are presented as an extraction context which corresponds to transaction-items relationship that are exploited by data mining algorithms. In our case, the extraction context expresses the access patterns of queries to attributes. Potential indexable attributes are structured in a text file where each row represents a query q_i and each column an indexable attribute (a_j) involved in the corresponding query. Each row i corresponding to the query q_i is duplicated f_i times which corresponds to the frequency of the query q_i. This ensures that we have good information about the attributes the queries are referecing. Figure 1 illustrates an example of query characteristics. The corresponding extraction context is presented in Figure 2.

				a	b	c
				a	b	e
				a	b	e
Query	Frequency	Attributes		a	b	e
q_1	1	a b c		a	d	
q_2	3	a b e		a	d	
q_3	2	a d		b	c	
q_4	2	b c		b	c	

Fig. 1 Example of queries characteristics   **Fig. 2** Extraction context

3 Related Work

Selecting the indexes appropriate for a workload involves searching a potentially very large space of different candidate configurations. Searching the space of alternative configurations is impractical since the problem is classified as NP-Hard [8]. Therefore, most proposed approaches [8][9][10][11][12][13] are based on greedy heuristics that prune the search space of potential useful indexes. In mono-table context, the common methodology for choosing indexes go through two steps [13]: (1) For each query in the workload, propose a set of candidate indexes that potentially benefits the query, (2) Choose a final optimized configuration from the candidate indexes in (1). Mostly, the choice of the final configuration relies on greedy search algorithms (bottom-up or top-down search) while the storage space constraint is considered.The bottom-up search strategies begin with an empty configuration and greedily add new indexes. On the other hand, the top-down search strategies begin with an optimal configuration which is progressively refined to be in accordance with storage constraint [14].

Due to the changes in query environments in data warehouses, new index structures are introduced. The most commonly used structures are bitmap join indexes. They are multi-attribute indexes involving several tables. As part of a research on the application of different data mining techniques for optimizing the physical design of decision systems, we have presented in [15] a solution to the studied problem based on mining maximal frequent itemsets. It leverages and extends principled reported methods for the bitmap join index selection problem using frequent itemsets technique [16][17]. Although the proposed techniques can alleviate the search space size by eliminating attribute groups that occur infrequently in the workload, they can fail their primary goal that is to select the most relevant and significant indexes, since they often lead to a loss of accuracy on the attributes correlation. Therefore we should not expect strong relevance of the mined indexes. For a much more effective pruning, we exploit the association rules concept. We believe that association rules constitute a powerful tool that can be applied to produce more useful configuration. Intuitively, the attributes correlation accuracy is improved using association rules rather than using the traditional frequents itemsets.

4 Association Rules Based Approach for Index Selection

We briefly describe our method for selecting bitmap join indexes. It takes as input a workload, a storage budget, support and confidence values and provides a set of indexes which can improve the performance of the given workload. The basic steps of the proposed approach are the following:

Construction of the extraction context: This step examines the input workload and derives the queries references to attributes in the form of an extraction context. Potential indexable attributes are structured in a text file where each row represents a query q_i and each column an indexable attribute (a_j) involved in the corresponding query. Each row i corresponding to the query q_i is duplicated f_i times which corresponds to the frequency of the query q_i. This ensures that we have good information about the attributes the queries are referecing.

Candidate configuration Selection: This step exploits the extraction context and eliminates the indexes (sets of attributes) that occur infrequently in the workload. To improve the accuracy of the selected indexes, we deduce a set of useful index configurations using association rules technique. A pruning process is necessary to elimminate redundant indexes. Finally, a generated configuration is eliminated if its space requirements violates the storage budget. The output of this processing is a set of candidate configurations.

Selecting the optimized configuration: Exploiting the configurations generated in the previous step, we evaluate the workload cost for each configuration and we pick the configuration leading to the minimum workload cost.

To illustrate the working of our approach, let us see a practical example. Consider the workload example given in Section 2. Figure 3 illustrates the extraction context for our example. We aim at selecting a reduced set of correlated attributes which will be used for the indexing procedure. Assuming, for example, a minimum support value of 25%, frequent indexes are presented in Figure 4. Corresponding association rules are summarized in Figure 5. Let $c = 60\%$ be the value of the confidence. A potential configuration is then $\{\{a,b\},\{a,d\},\{a,e\},\{b,c\},\{b,e\},\{a,b,e\}\}$. Finally, a redundancy elimination leads to the configuration $\{\{a,d\},\{b,c\},\{a,b,e\}\}$.

```
a  b  c
a  b  e
a  b  e
a  b  e
a  d
a  d
b  c
b  c
```

Fig. 3 Extraction context

Index	Support	Index	Support	Index	Support
a	$\frac{6}{8}$	a b	$\frac{4}{8}$	a b e	$\frac{3}{8}$
b	$\frac{6}{8}$	a d	$\frac{2}{8}$	-	-
c	$\frac{3}{8}$	a e	$\frac{3}{8}$	-	-
d	$\frac{2}{8}$	b c	$\frac{3}{8}$	-	-
e	$\frac{3}{8}$	b e	$\frac{3}{8}$	-	-

Fig. 4 Selected indexes (support=25%)

Rule	Confidence	Rule	Confidence	Rule	Confidence
$a \rightarrow b$	$\frac{4}{6}$	$e \rightarrow a$	$\frac{3}{3}$	$a \rightarrow b, e$	$\frac{3}{6}$
$b \rightarrow a$	$\frac{4}{6}$	$b \rightarrow c$	$\frac{2}{6}$	$b \rightarrow a, e$	$\frac{3}{6}$
$a \rightarrow d$	$\frac{2}{6}$	$c \rightarrow b$	$\frac{2}{3}$	$e \rightarrow a, b$	$\frac{3}{3}$
$d \rightarrow a$	$\frac{2}{2}$	$b \rightarrow e$	$\frac{3}{6}$	$a, b \rightarrow e$	$\frac{3}{4}$
$a \rightarrow e$	$\frac{3}{6}$	$e \rightarrow b$	$\frac{3}{3}$	$a, e \rightarrow b$	$\frac{3}{3}$

Fig. 5 Corresponding association rules

5 Experimental Evaluation and Preliminary Results

In this section, we report our preliminary experimental results on the performance
of our approach as compared with our previous work presented in [15]. In this study
we are more interested in usefulness accuracy of the generated indexes. Thus, our
comparaisons are performed with no restrictions on available disk space. All experiments were carried out on a PC with 3.4 Ghz Intel(R) Xenon(TM) and 1024 Mb
of memory running Linux Ubuntu 10.04. We have used the APB-1 Benchmark of
the OLAP Council [18], which simulates an OLAP business situation. It consists of
a fact table *Actvars* and four dimension tables *ProLevel*, *TimeLevel*, *CustLevel*, and
ChanLevel. The characteristics of used tables are resumed in Figure 6. We have used
a workload consisting of 60 star join queries involving 12 indexable attributes. To
evaluate the workload cost and the indexes storage requirements, we have used the
same theoretical cost models as in closely related works which are originally proposed in [16]. To generate desired indexes using association rules, we have used the
weka[1] tool. We have also implemented in Java an application which estimates the
interestingness of the generated indexes. It prunes the weka output removing redundant indexes. Finally, it measures the storage space for a generated configuration and
estimates the workload cost exploting that configuration. For comparative purposes,
we keep the same *support* value leading to the best performances in [15] which
is 5%. We then generate index configurations using several predefined *confidence*
values. Obtained results are summarized in Figure 7 where $Size(C)$ indicates the
storage space needed for storing the configuration C and $Cost(W,C)$ expresses the
workload cost exploiting the configuration C. We can note that the performance cost
increases along with the increase of the *confidence* threshold value. The reason is
that for high values of the *confidence*, the number of indexes decreases significantly
resulting in a high workload cost and a low strorage space requirements. The best
performance is obtained for *confidence* values of 10, 20, and 30%. Figure 8 summarizes the best performances achieved using the proposed approach compared to
a previous method which is based on mining maximal frequent itemsets. The new
approach leads to a cost improvement of 38.41% and a space saving of 13.03%.

---

[1] www.cs.waikato.ac.nz/ml/weka/

Table	# of Records	Size of records
Actvars	24786000	74
ProdLevel	9	24
TimeLevel	900	24
CustLevel	9000	72
ChanLevel	24	36

Fig. 6 Tables characteristics

Confidence[%]	10	20	30	40	50	60	70	80	90
$Size(C)[Go]$	7.21	7.21	7.21	6.63	6.63	6.01	5.71	5.71	5.71
$Cost(W,C)[Millions\ I/O]$	5.45	5.45	5.45	5.48	5.54	5.80	6.78	6.78	6.78

Fig. 7 Storage space and Workoad cost vs Confidence threshold

Technqiue	$Size(C)[Go]$	$Cost(W,C)[Millions\ I/O]$
Maximal frequent itemsets[support=5%]	8.29	8.54
Association rules[support=5%, confidence=10%]	7.21	5.45

Fig. 8 Maximal frequent indexes vs Association rules indexes

6 Conclusion

This paper addresses a basic issue in data warehouse physical design by proposing an approach to automaticaly select an optimized index configuration to improve the performance of a given set of star join queries using the concept of association rules. Although previous proposed techniques, that are based on mining frequent itemsets, can alleviate the search space size by eliminating attribute groups that occur infrequently in the workload, they still suffer from the less accuracy of the selected indexes and may lead to biased recommendations. The proposed approach was tested using the APB-1 benchmark. Experimental evaluation show that our approach leads to better perormances both for the workload costs and the storage space required for the generated configurations. As a perspective, we plan to adapt our work to handle other problem in physical database design such as partitioning and view materialization.

References

1. Ranjan, J.: Business Intelligence: Concepts, Components, Techniques and Benefits. Journal of Theoretical and Applied Information Technology 9(1), 60–70 (2009)
2. Inmon, W.: Building the Data Warehouse, 2nd edn. John Wiley & Sons, Inc., New York (2002)

3. Kimball, R., Ross, M.: The Data Warehouse Toolkit: The Complete Guide to Dimensional Modeling, 2nd edn. John Wiley & Sons, Inc., New York (2007)
4. Bornaz, L.: Optimized Data Indexing Algorithms for OLAP Systems. Database Systems Journal 1(2), 17–26 (2010)
5. O'Neil, P.E., Graefe, G.: Multi-table joins through bitmapped join indexes. SIGMOD Records 24(3) (1997)
6. Bruno, N., Chaudhuri, S.: Automatic physical database tuning: a relaxation-based approach. In: Proceedings of the SIGMOD Conference (2005)
7. Agrawal, R., Imielinski, T., Swami, A.N.: Mining Association Rules between Sets of Items in Large Databases. In: Proceedings of the 1993 ACM SIGMOD International Conference on Management of Data, pp. 207–216 (1993)
8. Chaudhuri, S., Datar, M., Narasayya, V.: Index Selection for Databases: A Hardness Study and a Principled Heuristic Solution. IEEE Trans. Knowl. Data Eng. 26, 1313–1323 (2004)
9. Agrawal, S., Chaudhuri, S., Narasayya, V.: Automated Selection of Materialized Views and indexes in SQL Databases. In: VLDB, pp. 496–505 (2000)
10. Chaudhuri, S., Narasayya, V.: An Efficient Cost-Driven index Selection Tool for Microsoft SQL Server. In: Proceedings of 23rd International Conference on Very Large Data Bases, Athens, Greece, pp. 146–155 (1997)
11. Feldman, Y.A., Reouven, J.: A knowledge based approach for index selection in relational databases. Expert Syst. Appl. 25, 15–37 (2003)
12. Chaudhuri, S., Narasayya, V.: Microsoft Index Tuning Wizard for SQL Server 7.0. In: ACM SIGMOD International Conference on Management of Data, pp. 553–554 (1998)
13. Chaudhuri, S., Narasayya, V.: Index merging. In: Proceedings of the International Conference on Data Engineering (ICDE) (1999)
14. Bruno, N., Chaudhuri, S.: Automatic physical database tuning: a relaxation-based approach. In: Proceedings of the SIGMOD Conference (2005)
15. Ziani, B., Ouinten, Y.: Combining Data Mining Technique and Query Frequencies for Automatic Selection of Indexes in Data Warehouses. In: Proceedings of the Tenth International Baltic Conference on Databases and Information Systems, Vilnius, Lithuania (2012)
16. Aouiche, K., Darmont, J.: Data Mining-based Materialized View and Index Selection in Data Warehouses. Journal of Intelligent Information Systems 33(1), 65–93 (2009)
17. Bellatreche, L., Missaoui, R., Necir, H., Drias, H.: Selection and pruning algorithms for bitmap index selection problem using data mining. In: Song, I.-Y., Eder, J., Nguyen, T.M. (eds.) DaWaK 2007. LNCS, vol. 4654, pp. 221–230. Springer, Heidelberg (2007)
18. Olap Council.: APB-1 Benchmark, http://www.olapcouncil.org/

The Utilization of the Moodle E-Learning System in Isra University

Mohammad Ali Eljinini, Zahraa Muhsen, Adi Maaita, Ayman Alnsour,
Mohammad Ali Azzam, and Khalil Ali Barhoum

Abstract. The Moodle e-learning system has been adopted by several universities
and organizations for the large accessible set of e-learning tools. Isra University
has adopted and utilized the Moodle e-learning system in many classes throughout
all of its colleges. Moodle has solved many course management problems such as
finding the appropriate time that suits all students to carry out tests and quizzes;
marking and providing feedback to the students within a short period of time; registration for tutorial sessions; and providing lecture materials and general faculty
announcements. Our study focuses on student survey to shed some light on the
perceptions of using Moodle in Isra University. The evaluation results of using
Moodle platform shows promising opportunities to support and improve upon this
platform in Isra University classes. This study helps introducing the e-learning
system to the entire students in Isra University and to support the understanding of
the overall learning process, learning motivation, legitimatize application of
knowledge, and a challenge for improving the teaching behaviors.

Keywords: Moodle, Open Source Software, Learning Management System,
E-learning, Virtual Learning Environment, Course Management System, Virtual
Learning comparison Environment.

1 Introduction

Online interactions provide a large knowledge exchange on variety kinds of information exchanged between users. There are many software systems available
that serve as online learning systems. These systems are expressed by forms,
commercial and Open Source Software (OSS). Moodle has been adopted by several universities and organizations in the entire world because it offers a large ac-
cessibly set of tools, it does have many components that was developed without a
specific design documentation including its security services [8][11].

Mohammad Ali Eljinini · Zahraa Muhsen · Adi Maaita · Ayman Alnsour ·
Mohammad Ali Azzam · Khalil Ali Barhoum
Faculty of Information Technology, Isra University, Amman, Jordan
e-mail: Z_muhsin@ipu.edu.jo

J. Casillas et al. (Eds.): *Management Intelligent Systems*, AISC 220, pp. 75–81.
DOI: 10.1007/978-3-319-00569-0_10   © Springer International Publishing Switzerland 2013

Computer based learning is a way of learning in which knowledge delivered electronically to remote learners via computer networks. All LMS must provide a major functionality which must be helpful and useful for the user to access these systems such as introducing an online forums and message boards, online testing, email, chat rooms, picture languages, instant messaging, remote screen sharing or multilayer online games. In 1960 the Plato learning system (Programmed Logic for Automated Teaching Operation), was introduced by Illinois University. In 1997 an important change appears when the WbCt 1.0 and Blackboard were released; these two systems affected the LMSs concept. In 1998 Moodle was introduced as Modular Object- Oriented Dynamic Learning Environment soon it became one of the most common used systems among LMS and finally released in 2001. Moodle has an ability of tracking the learner's progress, which can be monitored by both teachers and learners. This fact implicitly includes both security and privacy threats and makes Moodle vulnerable system [10].

Moodle has a good architecture, implementation, interoperability, and internationalization, and also has a high strength of the community. Moodle is released as a free software licensed under the general public license (GPL). The cost of ownership is free and its accessibility is average. On the other hand, it has limitations, notably the lack of SCORM support, and its roles and permissions system is limited. It does not support document transformation. However, these limitations are going to be fixed, which are parts of the project road-map [5].

Nowadays, many higher education institutions use course management system (CMS), as a tool to assist delivering course materials to students, either as a supplement to courses delivered traditionally or as an entire course offered online [8][11]. A CMS system is an alternative teaching method that reduces methodology problems associated with managing large classes. Initiative to implement Moodle as an e-learning platform in IU, was first started during 2007, as part of the development program for entrancing the teaching and learning process.

Isra University offers computer skills course for undergraduates to be taken at the first semester of their four years study. Figure 1 illustrates the diagram of students enrolled in IU with the computer skills course. The students attend two hours of theoretical lectures and one hour of tutorial per week. Moodle was introduced to IU in early 2007. Although Moodle has not been utilized to its full potential yet, it is found that e-learning platform is able to assist the assigned course. Students are distributed over selected sections, and hence there are more than 10 classes of 20-45 students each with its own timetable. A team of instructors including a course coordinator are allocated to teach computer skills course. Managing such a course in this particular setting is rather cumbersome. The common problems experienced by the team includes: handing out course materials, student-instructor communications, and registration for tutorial sessions, scheduling and conducting common quizzes/tests. Furthermore, the manual grading adds an extra burden to the instructors and it is almost impossible to provide immediate results and feedback.

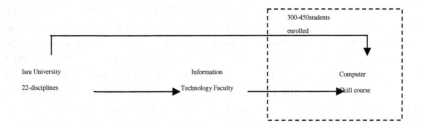

Fig. 1 Computer skill course in Isra University

This paper describes the experience in managing the CSC. A Moodle survey activity was conducted for the students' course to obtain feedback relating to the features used in Moodle and the management of the course. The paper is organized as follows: Firstly in section 2, the features of e-learning are explained briefly. Evaluating the impact of e-learning is shown in section 3. Section 4 presents the results of the survey. Finally, the conclusions are presented in section 5.

2   Feature of the E-Learning System

LMSs have many features and abilities expected from an e-learning system, such as forums, content management, quizzes with different kinds of questions, and a number of activity modules [16]. Table 1 shows, the features and capabilities provided by the Moodle system. Al-Ajlan and Zedan [1] have divided these features and capabilities into three parts, which are Learner Tools, Support Tools and Technical Tools, as in Table 1 [5][6][10], each kind of these tools holds minor tools as shown in Table 1.

Table 1 Features and capabilities hold in Moodle system

Major Tools	Minor Tools holds	Product name Tools
Learning	Communication, Productivity and Student Involvement	Discussion Forums, Discussion Management, File exchange, Internet Email, Online Journal, Real Time Chat, Video Services, Whiteboard, Bookmarks, Calendar, Orientation, Searching Course, Work off line, Group Work, Community, Student Portfolios.
Support	Administration, Course Delivery, and Content Development	Authentication, authorization, file exchange, registration integration, Test Types, Automated support, course management, online grading, student tracking, accessibility, Content Sharing, Course templates, Look and Feel, Design, Instructional Standards
Technical Specification	Hardware/Software and Pricing/Licensing	Client request, database requirements, Unix server, Windows, Costs, Open source, Optional extras

Moodle is a tool that is used all over the world by over 400,000 registered users, it is an Open Source Software (OSS) e-learning platform, which also features a community web site that provides developer information, roadmap, coding guide and Concurrent Versioning System (CVS) guide to access its source code, and it has a long list of developers. It does not provide a formal model for future development [10]. It is a tool that teachers can use to facilitate student collaboration in many ways, it is a platform to save and provide teaching material easily and a collaborative online platform for teachers and students to undertake the learning process together. Many modules can easily be found in the Moodle.org community website that is very useful to enrich the Moodle e-learning system with video illustrative materials. These tools help making the teaching process more effective. There are thousands of Moodle systems worldwide ranging from a single-teacher site to a 40,000-student university site [8][9][12][2][4][3][7][15]. Nevertheless, many Universities in Jordan that has adopted commercial CSM (WebCT or Blackboard) have considered using Moodle as an alternative.

3   Evaluating the Impacts of E-Learning

The most evaluating impacts for e-learning are both exploring issues related to the evaluation of the e-learning process and the teaching activities, and propos a comprehensive plan for the evaluation of e-learning and teaching. The key tendency of any evaluation activity is to influence decision making. This task can be completed by, a comprehensive evaluation strategy for ascertaining the impact of its various teaching; learning and research related activities are crucial. Any strategy required to gather different types of data and feedback and it also crucial to ensuring a high quality of service, and effective utilization of information and communications technology in teaching and learning [13][14].

The main difference of various types of educational evaluation activities are drawn between formative, and monitoring or integrative evaluation. Evaluation methodology depends on the data gathering process which should be simple and as less intrusive as possible, and they should be gathered from all students and staff regularly. The variety data gathering process also optional data from the following evaluation:

Front-end evaluation: this type of evaluation requires a surveys to ensure the desired and preferences in relation to teaching and learning online. *Formative evaluation*: this type of evaluation focuses on the feedback from users and other relevant groups during the implementation process. *Summative evaluation*: ensures the impact and outcomes of the e-learning process on teaching and learning at any university. Monitoring or integrative evaluation: this type of evaluation includes attempts to ensure the extent to which the use of e-learning or online learning is integrated into regular teaching and learning activities at any University.

4   Student Survey Result and Discussion

Moodle has great potential for creating a successful e-learning experience by providing a set of excellent tools that can be used to enhance the conventional classroom experience and turns it into an e-learning experience. In previous years many universities in Jordan have used Blackboard and Claroline as a LMS in their education, but many of them have replaced these two LMS with the Moodle e-learning system and mostly in 2011.

In order to evaluate and analyze the effects of Moodle, the students were asked to complete a small survey near the end of the course. The survey was conducted by 86 undergraduate, and 57% of student is female. Students were asked to identify the following questions as explained and showed in tables 2,3,4.

Table 2 Question Survey and Results

Question	Result
Do you like using computers?	85% of student listed it as They like the use of computer
What is the main purpose of using computer?	3% fun, 55% education, and 42% information collection
Do you find the e-learning process useful?	77% agree, 5% not Sure and 18% disagreed
Do you see that the use of the e-learning system affected the style of your study?	56%agree, 8% not sure, 36% disagreed
Do you need a training for using e-learning system?	36% agree, 18% not sure and 46% disagreed
Do you find it difficult to download lectures from e-learning site?	25% agree, 19% not sure, 56% disagreed

Table 3 Elaborations on the positive feedbacks

Percent	Discussion	Resound
97%	Used computer for learning and collecting information	Many courses in the University require the use of computer in education.
77%	First time using Moodle	Some students are transferred from other universities
56%	E-learning system have changed the study style	Many courses use the tradition techniques blended with the e-learning system
46%	Moodle is a simple to deal with	Moodle is usually explained in the first lecture of the course

Table 4 Elaborations on the negative feedbacks

Student percent	Description	Resound
18%	Do not like to exchange the traditional teaching style with the e-learning system	Some students are not keen into changing their learning style.
36%	Do not agree that the e-learning system did change their style of studying	This type of student face difficulty in their learning
36%	Find that e-learning system in IU needs a training tutorials	Do not have any computer skill
25%	Face certain difficulties to download lectures from the e-learning system.	Mostly face download problems

5 Conclusion

Many researchers conclusions clearly indicated the advantages of the Moodle system over many other related systems. The questioner conducted indicating the importance of the presence of the Moodle system as an e-learning tool in Isra University. The uses of LMS in courses save a lot of time, also organize and manage different concurrent sections for a particular course. The Moodle LMS contributed to change the study style amongst students. Students believe that the e-learning system is a useful tool and an open class for the entire time; where an online interaction with teachers and perform live discussions anywhere and any time is possible to easily find the missing class materials. The classes materials can fully be loaded to the system and hence can be reached by students.

The survey covers a range of issues in an attempt to capture the important information of the use of the e-learning system in IU in order to analyze the impact of using the e-learning system. The students were asked to complete a small survey near the end of the course. The survey was conducted for 86 undergraduate students enrolled in pre-computer skill courses; it was setup using the Moodle platform survey.

References

1. Al-Ajlan, A., Zedan, H.: Why Moodle. In: 12th IEEE International Workshop on Future Trends of Distributed Computing Systems, pp. 58–64 (2008)
2. Carlson, P.A.: Work in Progress - Using a Course Management System in K-12 Education. In: 39th ASEE/IEEE Frontiers in Education Conference, San Antonio, TX, October 18-21 (2009)
3. Bower, M., Wittmann, M.: Pre-service teachers' perceptions of LAMS and Moodle as learning design technologies. In: The 4th International LAMS and Learning Design Conference, Sydney, Australia, pp. 28–39 (2009)
4. Dobrzaski, L.A., Brytan, Z., Brom, F.: Use of e-learning in teaching Fundamentals of Materials Science. Journal of Achievements in Materials and Manufacturing Engineering 24(2), 215–218 (2007)

5. Dougiamas, M.: Moodle (2008), `http://www.Moodle.org`
6. EduTools. Course Management Systems (2007), `http://www.edutools.info/`
7. Gorghiu, G., Gorghiu, L.M., Suduc, A.M., Bîzoi, M., Dumitrescu, C., Olteanu, R.L.: Related Aspects to the Pedagogical Use of Virtual Experiments, Research, Reflections and Innovations in Integrating ICT in Education, Lisbon, Portugal, vol. 2, pp. 809–813 (2009)
8. Herdiana, A.: Moodle: Tool to Manage Probability and Statistics Course in Universiti Teknologi PETRONAS R. In: ICEE 2008, Hungary, July 27-31 (2008)
9. Herdon, M., Lengyel, P.: Multimedia and e-Learning integration for supporting training programs in agriculture by Moodle. In: AWICTSAE 2008 Workshop, Alexandroupolis, Greece (2008)
10. Kumar, S.A., Gankotiya, K., Dutta, K.: A comparative study of moodle with other e-learning systems. In: 3rd International Conference on Electronics Computer Technology (ICECT), Kanyakumari, pp. 414–418 (April 2011)
11. Machado, M., Tao, E.: Blackboard vs. Moodle: Comparing user experience of Learning Management Systems. In: Proceedings of the 37th ASEE/IEEE Frontiers in Education Conference, pp. S4J-7–S4J-12. IEEE (retrieved March 23, 2009)
12. Sinka, R., Papp, G., Vágvölgyi, C.: Open source information society from beginners to advanced' in the Hungarian education The possible roles of Moodle in the Hungarian teacher training Robert. In: ICL 2007, Villach, Austria, September 26 -28 (2007)
13. Som, N.: E-Learning: A Guidebook of Principles, Procedures and Practices (2003)
14. Sumak, B., Hericko, M., Pusnik, M., Polancic, G.: Factors Affecting Acceptance and Use of Moodle: An Empirical Study Based on TAM. An International Journal of Computing and Informatics 35(1), 91–100 (2011)
15. Weng, T., Lin, H.C.: The Study of E-Learning for Geographic Information Curriculum in Higher Education. Applied Computer Science (EI), 618–623 (2007)
16. Zenha-Rela, M., Carvalho, R.: Work in Progress: Self Evaluation Through Monitored Peer Review Using the Moodle Platform. In: 6th Annual Frontiers in Education Conference, pp. 230–241. IEEE, San Diego (2006)

Finding People Who Can Contribute to Learning Activities: A First Approach to Enhance the Information about Experts Available in a People Directory

Víctor M. Alonso Rorís, Agustín Cañas Rodríguez, Juan M. Santos Gago,
Luís E. Anido Rifón, and Manuel J. Fernández Iglesias

Abstract. The participation of experts and other external contributors is a
common requirement in the design of educational scenarios for the school of
the future. We can find many repositories of learning objects, but it is not so
common to find directories containing people who are willing to participate
in an educational activity. Much less common is to find information about
these people to determine their suitability from an educational perspective.
This paper describes a proposal for the automatic enrichment of existing information in a directory of contributors to educational activities. Through
this enrichment process, it is possible to enhance the amount of information
available to a recommender system to identify the most appropriate people
to participate in a particular activity, and reduces the need for human intervention when selecting individuals to contribute to educational activities.

Keywords: RDF, Record Linkage, Enrichment, Semantic Web.

1 Introduction

Education, as it is presently understood, must evolve towards a more open
model in which educators can receive support and assistance from external
experts to develop their educational initiatives. This vision is included among
the objectives of several projects in the field of technology-enhanced learning.
In our case, we are developing, in collaboration with renowned educational
institutions, an intelligent system to support teachers when planning and
developing their educational activities.

Víctor M. Alonso Rorís · Agustín Cañas Rodríguez · Juan M. Santos Gago ·
Luís E. Anido Rifón · Manuel J. Fernández Iglesias
Telematic Systems Engineering Group, Universidade de Vigo, Vigo, Spain
e-mail: {valonso,agustincanas,jsgago,lanido,manolo}@det.uvigo.es

J. Casillas et al. (Eds.): *Management Intelligent Systems*, AISC 220, pp. 83–90.
DOI: 10.1007/978-3-319-00569-0_11   © Springer International Publishing Switzerland 2013

The amount of educational resources available and the complexity and variety of educational settings may pose a challenge for most present educators. In order to reduce this complexity, we have designed an intelligent system capable of recommending educational resources based on the specific context of each educational activity. These resources include experts available to carry out learning experiences. One of the key factors that determines the accuracy of any recommender system is the accuracy and completeness of the information available in its knowledge base (KB). In most cases, this information is generated manually by the user community, teachers in many cases. Unfortunately, this process is limited by the cataloguer's prior knowledge.

On the other side, we can find on the Web hundreds of sources with detailed and accessible information about the same educational resources users are expected to generate. Based on this perspective, this paper discusses a mechanism to supplement the information provided by the cataloguers using several information sources available on the Web.

This paper discusses how to enrich using external sources the semantic descriptions of experts registered in a preexisting directory. We use as a case study the work performed in the framework of the iTEC project, an European 7th Framework Programme project. We introduce in Sect. 2 the iTEC expert recommendation system. In Sect. 3 we describe the types of external sources considered, and mechanisms used as a proof of concept of the enrichment process. Sect. 4 discusses the first results, and finally Sect. 5 provides some conclusions drawn during the development of this project.

2   Recommender System

The system collects descriptions provided by experts according to a semantic model expressed in OWL [5] (cf. Fig. 1).This model defines a collection of properties that enable the system to automatically provide smart recommendations from complex queries by means of semantic rules and inference engines. More specifically, it will provide lists of experts that can satisfy the requirements of the educator. This list will be sorted according to several factors. Among them, we can highlight the factors below:

- Expertise: Subject field in which the expert excels. This factor is weighted according to the degree of expertise.
- Contact: Information that enables to get in touch with the expert (e.g., e-mail, phone, social network profile, etc.).
- Location: Geographical areas where the expert will be able to physically participate in the educational activity (e.g., job or private address, etc.).
- Languages: Languages spoken by the expert.
- Knows: Collects relations among experts.
- MemberOf: Links experts to associations and groups (e.g. universities, companies, interest groups, etc.).

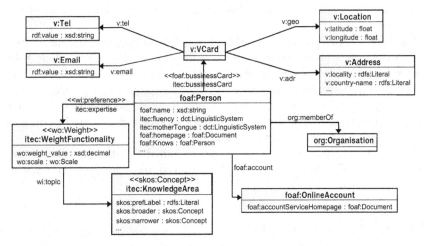

Fig. 1 Outline of the data model proposed to represent domain experts

The system will use these factors to sort the experts found according to their relevance or adequacy to satisfy the requirements established by the educational activity. This multi-criteria approach dramatically improves the quality of the results provided. This recommendation process is out of the scope of this paper. For more information, please refer [1].

To enable the system to exploit to the most the recommendation process, the knowledge base must include the most complete and accurate expert descriptions possible. In real situations, the user community inserts manually most of the information in the knowledge base. This manual process tends to be tedious and non-exhaustive, and as a consequence, descriptions provided use to be inaccurate and incomplete. Thus, an independent module has been developed, transparent to the recommender system, aimed at supplementing the initial knowledge in the KB with information freely available on the Web.

3 Enrichment Module

The enrichment module comprises a collection of agents that, according to some extraction patterns, obtain, process, convert and insert in the KB external information on the experts already in it. Agents and extraction patterns are designed and implemented to adapt to the formats in which the information sources represent the information they host.

The purpose of our proposal is to enrich the information of experts registered in the iTEC people directory. The main factor that determines at which extent an expert is suitable to a learning activity is the expert's fields of expertise. In many cases, knowledge areas mastered by a given person are reflected in their scientific publications. As a consequence, relevant sources for enrichment are online portals including information on academic publications.

However, to determine a suitable recommendation ranking, the system must take into account other aspects besides mastering a particular field. For example, it is also relevant whether the expert can communicate in the language of a given educational activity, whether the expert is physically close to the venue of the educational activity, etc. Relevant sources to supplement this information are social networks' profiles, as they use to share information on language, location, contact details, etc.

We describe below some of the sources studied and the enrichment mechanisms used. For this, we will introduce each source according to the nature of the information exposed.

3.1 HTML-Based Sources

This type of source is the most common on the Internet and is specifically designed to be human-friendly. On the other side, information extraction using software agents is a complex task due to the lack of a rigid structure in HTML documents. The developed software agent acts as a web scrapper[3] based on an extraction pattern that analyses the DOM tree of each source and extracts the information from the tags identified. This information is adapted to the vocabulary of the existing data model in our directory to be inserted into the KB.

In this scenario we have developed a software agent and extraction patterns specific to Google Scholar (http://scholar.google.com). Google Scholar is a scientific search service in which users may create their own profile with their interests and academic knowledge, the institution for which they work and references to academic contributions, among other information.

Google Scholar is a service offered through HTML pages. Therefore, the enrichment agent submits queries to find profiles through keywords by encoding these keywords into the search URL. In this case, the name of the expert to be enriched is used. When querying for a person, the search returns an HTML page that lists the names of users similar to the one in the query. The agent compares the name with each of the names returned using the Jaro [4] syntactic matching heuristic and keeps only those that exceed a similarity threshold of 0.97. In this way, it is possible to detect profiles that reference the target user and discard those that are not directly related. This process of linking records that refer to the same object in different sources is known as record linkage [7] and is the basis of the enrichment process.

As a practical example, let us take the expert named Victor Manuel Alonso Roris. As a result of a Google Scholar query, the system returns a single profile, which is inferred to be valid because the name syntactically matches the target name. Through the returned profile, the system finds out that the expert: 1) is a member of the University of Vigo, 2) has several knowledge areas, as posted in his profile (i.e., Semantic Enrichment, Linked Data, etc.), and 3) has published a collection of papers. Based on these articles, the system may obtain further areas of expertise, and infer relations with other experts in the KB through the papers' co-authors.

It should be pointed out that HTML sources require in many cases active participation from humans to design the complex extraction patterns used by the enrichment agent. Besides, these extraction patterns depend on the DOM tree structure. Any change to it, like the one consequence of a new layout for the result page, will deem the extraction pattern invalid.

3.2 XML-Based Sources

XML sources present information in a structured way that facilitates its automatic processing. However, the layout and rendering of each XML scheme is proprietary to the corresponding provider. As a consequence, a specific extraction pattern has to be designed for each XML source. In any case, patterns are much simpler than HTML extraction patterns, and XML schemes are more robust and less prone to change.

As a general rule, XML sources are accessible through application programmer's interfaces (API). In our case, we have developed an agent to access the LinkedIn API. LinkedIn (http://www.linkedin.com)is a professional social network in which each community member publishes his academic and professional profile including studies, job, location, language, etc. Many of these fields are public and accessible through the API.

Similar to the process described in the previous section, the agent developed queries the target source according to the name of the expert to enrich. As a response to this query, the agent receives an XML-formatted list of users that match the desired parameters. Then, record linkage is performed through the syntactic comparison of each name in the list, to eventually recover the matching profiles.

As an illustration, let us perform the enrichment of expert Juan Manuel Santos Gago using LinkedIn. We use the API to perform a query using the expert's first and last name. As a response, an XML-formatted profile list is returned, where the first element in the list has the greatest possible similarity to the submitted name. Therefore, the agent infers that this profile corresponds to the expert to enrich. Using the identifier associated with this element, the agent performs additional queries through the API to discover and retrieve the public fields in the profile. Thus, through this enrichment process it was possible to achieve: 1) the location of the expert in Vigo, Spain, 2) its affiliation to the University of Vigo, 3) the skills or fields of expertise that he has published (e.g., e-learning, Semantic Technologies, etc.), and 4) his fluency in English, Spanish and Galician.

3.3 RDF-Based Sources

RDF-based sources are designed to make it accessible to software agents large amounts of interconnected information, and more specifically pieces of information that follow the design principles proposed by the Linked Open Data initiative (LOD [2]). For these sources, there are tools that support fully automated record linkage processes according to previously defined extraction patterns. This is the case of SILK [6], a tool that performs the matching process and returns a list of records identified as equal according to a configuration file defining a collection of factors (e.g., sources, entities and properties, heuristics, etc.).

In the case of experts, there is a collection of LOD nodes that record relevant information about authors of scientific publications, and experts in general. Among these, DBLP semantic nodes (http://dblp.rkbexplorer.com) and DBPedia (http://dbpedia.com) are particularly relevant. In general, the data models of these repositories tend to use the same vocabularies to identify entities that refer to people (usually foaf:Person and akt:Person) and to express their personal information (e.g., foaf:name and akt:full-name). This enables SILK to process all these semantic repositories according to a standard configuration. Listing 1 depicts an excerpt of the configuration file Mentioned above. Line 12 in that file defines the property used to link similar records (owl:sameAs in our case); sentences between lines 13 and 25 define concepts in each repository to be compared (in our case, concepts typically defining experts); sentences between lines 28 and 37 define the properties to be analysed for each concept, together with the comparison heuristic (in our case, values from properties foaf:name and akt:full-name, using the Jaro syntactic similarity metric); and finally sentence in line 59 defines the degree of similarity to consider two record as representing the same person (0.96 in our case).

To demonstrate this mode of operation, SILK has been launched on DBLP according to the configuration described above. The goal is to enrich the information on expert Luis E. Anido Rifón. As a result, we have obtained the URI of the entity that represents this expert in the repository. Using it, the foaf: relations made it possible to navigate among the papers associated with this expert (presently a reference to more than one hundred articles). Based on this process, it was recovered for each item: 1) title, 2) co-authors, and 3) the home page address (property akt:has-web-address) of the article. Most of these papers are indexed in a limited number of sites. This has made it possible to develop a collection of very simple Web scrappers to extract the keywords of each paper. Assuming that keywords represent areas mastered by the expert, the corresponding knowledge areas of Luis Anido can be enriched with new references (e.g., educational technology, etc.).

Listing 1 Excerpt from the SILK configuration file for comparing two different datasets according to their properties

```
12: <LinkType>owl:sameAs</LinkType>
13: <SourceDataset dataSource="external" var="a">
14:    <RestrictTo>
15:       {?a rdf:type akt:Person} UNION {?a rdf:type foaf:Person}
16:    </RestrictTo>
17: </SourceDataset>
18: <SourceDataset dataSource="local" var="b">
19:    <RestrictTo>
20:       ?b rdf:type foaf:Person
21:    </RestrictTo>
22: </SourceDataset>
...
28: <Aggregate type="max">
29:    <Compare metric="jaroSimilarity">
30:       <Input path="?a/foaf:name" />
31:       <Input path="?b/foaf:name" />
32:    </Compare>
33:    <Compare metric="jaroSimilarity">
34:       <Input path="?a/foaf:name" />
35:       <Input path="?b/akt:full-name" />
36:    </Compare>
37: </Aggregate>
...
59: <Output minConfidence="0.96" type="file">
...
```

4   Discussion

In the current state of development of the enrichment module, it must be considered a proof of concept based on simple mechanisms for extracting information. Through this process we have obtained feasible semantic descriptions to enrich and improve the recommendations in the iTEC project. In relation to the processes described in the previous sections, the system still encounters some problems that limit the success of the enrichment strategy, namely 1) the detection of specific iTEC fields of expertise, and 2) the measurement of the degree to which an expert masters a given field of expertise.

For the detection of fields of expertise, the problem stems from the fact the driving iTEC vocabulary is limited to only 37 fields of expertise, which doesn't cover all areas that can be found in Web profiles. To circumvent this limitation, the enrichment module applies further record linkage techniques to find terms related to the ones in the vocabulary (i.e., synonyms, generalizations, concretions, etc.). Among the repositories studied we selected 1) WordNet (http://wordnet.rkbexplorer.com/), a lexical database that provides definitions for terms, 2) DBPedia, which includes a broad hierarchical classification of universal concepts.

Once a given field of expertise has been detected for an expert, the system must compute the degree to which the user masters the corresponding subject. In our implementation, this weight increases according to the number of relations between expert and the field in the Web.

5 Conclusions

The enrichment methodology described in this paper can be applied to complement any semantic description, and not just experts. The success of the process is determined by the amount and quality of knowledge freely shared on the Web in relation to the subject to be enriched. Besides, another key factor limiting the enrichment performance is the nature of the information available, mainly due to its level of structure, which determines the amount of effort required to generate more or less complex extraction patterns.

In any case, it should be noted that the enrichment process is made not only during the information extraction stage, but then data must be processed and conditioned to match the data models and the limitations of each specific system. The latter task requires active user participation in the design of conversion algorithms and patterns.

Acknowledgements. This research is funded by the European Commision's FP7 programme - project *iTEC: innovative Technologies for an Engaging Classroom* (Grant nr. 257566) and the Galician Government through grants *Research Networks: TELGalicia* (CN 2012/259) and *Red PLIR* (CN 2012/319).

The authors of this paper is solely responsible for it and it does not represent the opinion of European Commission or the Galician Government.

References

1. Anido, L., Caeiro, M., Cañas, A., Fernández, M., Míguez, R., Alonso, V., Santos, J.: D10.2 - Support for implementing iTEC Engaging Scenarios V2 (2012)
2. Bizer, C., Heath, T., Berners-Lee, T.: Linked Data - The Story So Far. Inter. Journal on Semantic Web and Information Systems 5(3), 1–22 (2009)
3. Ferrara, E., Fiumara, G., Baumgartner, R.: Web Data Extraction, Applications and Techniques: A Survey. ACM TOCL V, 1–20 (2010)
4. Jaro, M.A.: Probabilistic linkage of large public health data files. Statistics in Medicine 14(5-7), 491–498 (1995)
5. McGuinness, D.L., Van Harmelen, F.: OWL Web Ontology Language Overview. W3C recommendation 10, 1–22 (2004)
6. Volz, J., Bizer, C., Gaedke, M., Kobilarov, G.: Silk – A Link Discovery Framework for the Web of Data. Language 135(2), 1–6 (2009)
7. Winkler, W.E.: Overview of record linkage and current research directions. Current Statistics (2006-2), 1–28 (2006)

Providing Event Recommendations in Educational Scenarios

Agustín Cañas Rodríguez, Víctor M. Alonso Rorís, Juan M. Santos Gago,
Luís E. Anido Rifón, and Manuel J. Fernández Iglesias

Abstract. This paper introduces a novel event recommendation system for
educational scenarios. Unlike traditional recommender systems that base
their recommendations on user feedback, the proposed system takes into
account both existing information on events and the particularities of the
specific target learning environment. For this purpose, the system addresses
the problem of recommendation as a decision problem based on multiple criteria. As part of the developed system, we propose a collection of interest
criteria and an algorithm to sort the events according to their relevance to
the development of a given learning activity. This work is the result of the
collaboration between technical and pedagogical experts in the framework of
an European large-scale pilot focused on learning in the 21st century and
the design of the future classroom: *Innovative Technologies for an Engaging
Classroom (iTEC)*.

Keywords: Learning resources, Educational events, Recommender Systems,
Multi-Criteria Decision Analysis

1 Introduction

In spite of having more and more sophisticated tools available in European
schools (e.g. interactive blackboards, last generation learning management
systems, smartphones and tablets), mainstream education practice has not
evolved yet to be able to take the most of these new advances. Thus, the

Agustín Cañas Rodríguez · Víctor M. Alonso Rorís · Juan M. Santos Gago ·
Luís E. Anido Rifón · Manuel J. Fernández Iglesias
Telematic Systems Engineering Group, Universidade de Vigo, Vigo, Spain
e-mail: {agustincanas,valonso,jsgago,lanido,manolo}@det.uvigo.es

J. Casillas et al. (Eds.): *Management Intelligent Systems*, AISC 220, pp. 91–98.
DOI: 10.1007/978-3-319-00569-0_12   © Springer International Publishing Switzerland 2013

question posed in this context is "how can we take the most of new digital resources to better motivate students and improve pedagogical practice?"

Project iTEC [4] proposes a solution to this situation by defining as its main objective to develop engaging scenarios for learning in the future class-room that can be validated in a large-scale pilot, and be subsequently taken to scale. New scenarios proposed by iTEC pose a relevant challenge to teachers, as they are the ones that will eventually implement them. Due to the increasing number of educational resources (i.e. events, people, learning object, tools) that may be used to support learning activities, it is not easy for teachers to localize and select the best resources available to satisfy the requirements of educational activities, or even improve them.

As part of this project, we have created a recommendation system that takes into account the specific context where educational activities are performed to suggest the best educational resources to develop them. To accomplish this goal, the proposed recommender departs from the traditional user-centric recommenders' approach by focusing its recommendations in a specific learning context. Furthermore, to consider the special characteristics of these contexts, the design of the recommendation algorithm is based on multiple-criteria decision analysis (MCDA) techniques.

This paper discusses the developed system emphasizing the part that is responsible for the recommendation of events related to a given educational activity. The rest of the article is organised as follows: Sect. 2 provides some background information on recommender systems. Sect. 3 describes the proposed event recommender and presents initial results. Finally Sect. 4 provides the main conclusions of the first experiments, and some issues to deal with in the short term.

2 Background

Technology Enhanced Learning (TEL) aims to design, develop and test socio-technical innovations that will support and enhance learning practices of both individuals and organisations [6].

In recent years, some major TEL activities have focused on finding resources for the development of specific learning activities. As research advanced in this field, the problem of the availability of resources was losing prominence against the "findability" of the best resources for a particular learning activity. In these situation, the application of recommendation techniques may dramatically facilitate the learning and teaching processes. Indeed, in recent years the development of TEL recommender systems has attracted increased interest. [6] provide an extensive in-depth analysis of this kind of recommenders.

2.1 Recommender Systems

According to [10],

> "Recommender systems suggest items of interest to users based on their explicit and implicit preferences, the preferences of other users, and users and item attributes."

Traditionally, a recommender system use opinions of a community of users to help individuals in that community to more effectively identify content of interest from a potentially overwhelming set of options [5]. Formally, an recommendation system tries to estimate the rating function R from an initial set of ratings that is either explicitly provided by the users (opinions) or is implicitly inferred by the system. Let *Rating* a totally ordered set, *Users* the users' domain and *Items* the domain of candidate items to be recommended. The utility function R is defined by:

$$R : Users \times Items \rightarrow Rating \qquad (1)$$

According to the way recommendations are made, recommender systems have been classified into two main types according to their recommendation strategy, namely content-based recommendation and collaborative recommendation. Moreover, additional types have been proposed in the literature. Table 1 summarizes the most relevant.

Table 1 Recommendation Systems classification [6]

Content-based recommendation	Users are offered recommendations on items similar to the ones they have preferred in the past
Collaborative recommendation	Users are offered recommendations on items that people with similar tastes and preferences have liked in the past
Utility-based recommendation	Recommendations are made according to the computed utility of each item for a given user
Demographic recommendation	Users are classified according to the attributes of their personal profile, and recommendations are based on demographic classes
Knowledge-based recommendation	Items are suggested based on logical inferences about users' preferences
Hybrid recommendation	Combines two or more of the above approaches

2.2 Multicriteria Recommender Systems

Multiple-criteria decision analysis (MCDA) is a subdiscipline of operations research that explicitly considers multiple criteria in decision-making environments. Recommender systems that use MCDA techniques are named multicriteria recommender systems, and they take into account several rating factors to combine them to offer better recommendations. Generally speaking, they use an utility function that reduces the multiple-criteria problem to a single-criterion one.

Previous research in this field [8] suggests to follow Roy's MCDA methodology [9] to model multi-criteria recommendation problems. According this methodology, the analysis of the decision problem consists of four steps:

1. **Defining the object of decision**: defining the set of items upon which the decision has to be made.
2. **Defining a consistent family of criteria**: identifying and specifying a set of functions that declare the preferences of the decision maker.
3. **Developing a global preference model**: defining the function that synthesizes the partial preferences upon each criterion
4. **Selecting a decision support process**: design and development of the procedure, methods, or software systems that will support a decision maker

3 An Event Recommender System for Learning

As discussed in Sect. 2.1, traditional recommender systems focus on recommending the most relevant items to users. Moreover, they usually consider a single attribute of a given item. These systems define an utility function based on a two-dimension model: *Items* and *Users*. More recently, context-aware recommendation systems (CARS) deal with modelling and predicting users' tastes and preferences incorporating available contextual information into the recommendation process as explicit data categories [8]. In this case, they use a three-dimensional approach (i.e., $Items \times Users \times Context$) or even a multi-dimensional model considering different contextual information as additional dimensions (e.g., $Users \times Items \times Time \times Location \times ...$) [1].

Typically, participants in an educational setting have different profiles according to their role (e.g., teacher, learner, parent, expert, etc.), their level of knowledge or expertise, or their objectives, among other features. The items suggested by a traditional (user-centric) recommender system to actors in an educational scenario may conflict if they are focused on each of the different profiles. Since the main purpose of our system is to rank the best events for the development of a learning activity in a given educational scenario, the solution proposed in this paper presents a combination of both traditional RS and CARS approaches. Specifically, our recommender proposes a two-dimensional model by replacing the *Users* dimension with *Context* dimension ($Context \times Items$). The particular users' characteristics and/or preferences are shifted to the background and are used just to improve the characterization of the learning context or to improve events' descriptions (through users' reviews).

Assuming the aforementioned switch of the *Users* and *Context* dimensions, the proposed system could be classified as an utility-based one (cf. Table 1) because it makes suggestions based on the computation of the utility of an item (item) in a given context (learning context).

3.1 Learning Context and Event Models

Context is a very broad concept that has been studied across different research disciplines, including computer science, cognitive science or organizational sciences, among others. Looking for a formal definition, it can be stated that context is a set of circumstances that form the setting for an event, statement, or idea, and in terms of which it can be fully understood [7]. In our system, the context is a collection of settings for educational scenarios. More specifically, as part of the learning context we define several variables that are classified into three groups (cf. Table 2).

Table 2 iTEC Learning Context variables

Related to participants	age range, education level
Related to the learning activity	subject, language, completion dates
Related to the educational scenario	location, technical settings (available tools)

On the other hand, the iTEC project defines an event as something that happens or takes place at a determinable place and time, particularly one of importance [3]. Workshops, seminars, conferences and virtual meetings are examples of events that may support novel learning activities to improve the educational practice in European schools. According to this definition, and taking into account the aims of the recommendation system introduced in this paper, a set of properties relevant to a learning activity were identified, including age range, completion dates, cost, education level, language, organizer, place (e.g. aquarium, museum, zoo), required tools (e.g. Netmeeting), subject and type of event (e.g. seminar, conference).

3.2 Recommendation Operation: Relevance Calculation

The main purpose of the iTEC event recommender system is to suggest sorted lists of events most suitable to the particular needs of a learning activity in a specified learning context. To accomplish this objective, the recommender takes as input the requirements of the learning activity, and the learning context settings. Once all data are conveniently processed, the process of creating the lists above is organised into two different processes, namely filtering and relevance calculation:

1. **Filtering**: taking into account the requirements of the learning activity, the recommendation system filters out those events that do not meet these requirements. Typical reasons to exclude an event are celebration dates, used languages or the type of event. As result of this filtering process, a collection of potentially valid events is obtained.

2. **Relevance calculation**: the recommender computes a relevance value for filtered events using an algorithm constructed according to the multicriteria recommendation systems' guidelines (cf. Sect. 2.2). Specifically, the general methodology for modelling decision-based problems proposed by [9] was applied.

Equation 2, obtained as a result of the application of the methodology mentioned above (cf. Sect. 2.2), summarizes the algorithm used for relevance calculation. It represents an utility function that synthesizes the marginal utility functions concerning each of the factors (f_i) identified along the initial steps of the methodology. More specifically, we use a weighted sum (w_i) of marginal utility functions (f_i).

$$R = \sum_{i=0}^{n} w_i \cdot f_i \tag{2}$$

iTEC Control Boards are a group of experts with both pedagogical and technological expertise. This group of domain experts had a special relevance in the selection and evaluation of the factors proposed. On one hand, they were responsible for selecting the factors of interest to the recommendation of events. On the other hand, they evaluated the relevance of each selected factor. Specifically, 53 iTEC partners were surveyed to mark the importance of each factor on a scale $[1, 10]$.

To identify a collection of potential factors for event recommendation, we thoroughly analysed properties of both learning contexts and event resources. As discussed above, from this initial collection, iTEC experts (iTEC Control Boards) were asked to select those of interest to the recommender. Then we defined, for each factor, an utility function enabling its quantitative evaluation in the range $[1, 10]$. As general rule, this function will take value $f_i = 5$, if the factor was not evaluated due to the lack of information; $f_i = 1$ when the event does not fit at all to the target learning context according to this factor; and $1 \leq f_i \leq 10$ otherwise. In this way, we tried to penalize the less interesting events and minimize the impact of the lack of information. Table 3 describes each of the factors considered (f_i) in recommendation process.

Table 3 Event factors

Subject (f_0)	This factor is used to rate an event according to its thematic area(s), i.e. the knowledge area to which the event is related.
Rating (f_1)	The value associated with this rating factor may be explicitly indicated or it may be computed from collaborative ratings.
Cost (f_2)	The aim of this factor is to prioritize free events. According to this rule, free events will receive the maximum value.
Organization (f_3)	Models the relevance of the event's organizer as a numerical weighted value.
Location (f_4)	This factor indicates the degree of geographical proximity of an event to the location where the learning activity is performed.
Required tools (f_5)	The aim of this factor is to promote events that can be accessed using tools already available to perform the learning activity.
Age range (f_6)	The value of factor f_6 will prioritize events being explicitly targeted to an audience (specified for the learning context).
Education level (f_7)	This factor is intended to prioritize events being explicitly targeted to an educational level (defined for the learning context).

3.3 Initial Results

The proposed system has been initially tested through a testing prototype in two ways: (1) in the framework of iTEC workshops, iTEC Control Boards have been asked on its potential benefits. On this issue, most of respondent people consider that recommender has the potential to lead innovation in the classroom. In other hand, (2) from a example dataset of events described in a different ways, experts on RS and researchers in TEL systems have been asked on the quality of the recommendations. They consider that recommendations provided are qualitatively better as the quantity and quality of the available information increase. [2] presents a more detailed analysis of these first results.

4 Conclusions and Future Work

The work discussed in this paper proposes a solution for event recommendation in educational scenarios. More specifically, a system was developed that takes into account a specific context where educational activities are performed to suggest the best events to improve or facilitate their development. For this purpose, our recommender, unlike most RS found in the literature (i.e., user-centric RS), focuses its recommendations on a learning context as defined in the iTEC project. In regard to the design of the recommendation algorithm, we applied MCDA techniques that enable to consider different factors identified by the collaboration of a group of experts with both pedagogical and technological expertise.

After analysing the first results, and despite mechanisms have been introduced in the algorithm to minimize the impact of the lack of information, the results demonstrated that the recommendations provided are qualitatively better according to the amount available information on events. To improve the recommender's efficiency, the authors are working on the introduction of information enrichment strategies to automatically supplement the information on events provided by teachers and other users.

As the recommender system focuses on providing recommendations fitting a given educational context, and not suggestions directly targeted to final users, its evaluation is not a simple task. The authors are also collaborating with domain experts to define metrics besides users' perceptions that would enable us to conveniently evaluate recommendation results.

Acknowledgements. This research is funded by the European Commision's FP7 programme - project *iTEC: innovative Technologies for an Engaging Classroom* (Grant nr. 257566) and the Galician Government through grants *Red PLIR* (CN 2012/319) and *Research Networks: TELGalicia* (CN 2012/259).

The authors of this paper is solely responsible for it and it does not represent the opinion of European Commission or the Galician Government.

References

1. Adomavicius, G., Sankaranarayanan, R., Sen, S., Tuzhilin, A.: Incorporating contextual information in recommender systems using a multidimensional approach. ACM Trans. Inf. Syst. 23(1), 103–145 (2005)
2. Anido, L., Caeiro, M., Cañas, A., Fernández, M., Míguez, R., Alonso, V., Santos, J.: D10.2 - Support for implementing iTEC Engaging Scenarios V2 (2012)
3. Anido, L., Caeiro, M., Cañas, A., Fernández, M., Míguez, R., Santos, J.: D10.1 - Support for implementing iTEC Engaging Scenarios V1 (2011)
4. European Commission's FP7 Programme: iTEC Project (2011)
5. Herlocker, J.L., Konstan, J.A., Terveen, L.G., Riedl, J.T.: Evaluating collaborative filtering recommender systems. ACM Trans. Inf. Syst. 22, 5–53 (2004)
6. Manouselis, N., Drachsler, H., Vuorikari, R., Hummel, H.G.K., Koper, R.: Recommender Systems for Learning. Springer (2013)
7. Oxford English Dictionary A. Oxford University Press (2012)
8. Ricci, F., Rokach, L., Shapira, B., Kantor, P.B. (eds.): Recommender Systems Handbook. Springer (2011)
9. Roy, B.: Multicriteria Methodology for Decision Aiding. Nonconvex Optimization and Its Applications. Kluwer Academic Publishers (1996)
10. Schein, A., Popescul, A., Ungar, L., Pennock, D.: Croc: A new evaluation criterion for recommender systems. Electron. Commer. Res. 5, 51–74 (2005)

A New Generation of Learning Object Repositories Based on Cloud Computing

Fernando de la Prieta, Javier Bajo, Paula Andrea Rodríguez Marín,
and Néstor Darío Duque Méndez

Abstract. This work presents a proposal for an architecture based on a cloud computing paradigm that will permit the evolution of current learning resource repositories. This study presents current problems (heterogeneity, interoperability and low performance) of existing repositories, as well as how the proposed model will try to solve them.

Keywords: Learning Object Repository, Cloud Computing, eLearning.

1 Introduction

It has become common in recent years to encapsulate educational resources in the form of learning objects (LO), a process that facilitates their management and reuse. That is, the systematic management of learning resources makes their dissemination possible.

To facilitate these dissemination tasks, the LO are stored in educational repositories. However, these repositories present problems at a technical level, such as low-level performance, unavailability, security, reliability, etc. At the same time,

Fernando de la Prieta
Department of Computer Science and Automation Control, University of Salamanca.
Plaza de la Merced s/n, 37007, Salamanca, Spain
e-mail: fer@usal.es

Javier Bajo
Department of Artificial Intelligence. Techinical University of Madrid.
Campus Montegancedo, Boadilla del Monte, Madrid, Spain
e-mail: jbajo@fi.upm.es

Paula Andrea Rodríguez Marín · Néstor Darío Duque Méndez
National University of Colombia, Colombia
e-mail: {parodriguezma,ndduqueme}@unal.edu.co

J. Casillas et al. (Eds.): *Management Intelligent Systems*, AISC 220, pp. 99–106.
DOI: 10.1007/978-3-319-00569-0_13   © Springer International Publishing Switzerland 2013

the paradigm as a whole also shows many deficiencies like the existence of too many schemas of metadata or interoperability specifications, or even internal architecture of the repositories.

This environment, which is complex not only at the technical level but at the conceptual level as well, has led to a low level of implantation of this technology despite the great advantages derived from its use. Thus, the objective of this study is to propose and subsequently develop a new model which can make it possible to reduce or remove the existing problems. To this end, Cloud Computing paradigm is the key to offering effective and efficient services such as storage, and the search and retrieval of educational resources

The next section presents the state of the art within this context, as well as a study to demonstrate the problems observed in repositories. Section 3 presents the detail of the architecture and, finally, section 4 the conclusions and future work.

2 Learning Objects and Repositories

The encapsulation of education resources in the form of LO makes their reutilization possible. Many authors have recently been presenting their vision regarding this concept [8][4][7], which has led to the appearance of a number of definitions. The aim of this study is not, however, to establish what an LO is, rather to simply remark that there is a clear consensus that an LO must be the minimal reusable unit with a specific learning objective.

There is also a consensus that each LO has to be associated with an external structure of metadata. This metadata allows making a first approach to the educational resource. In other words, the metadata permits improving the utility of the resource, since it makes its retrieval, search, exchange, and hence, its reutilization, possible. The metadata schema is standardized. In fact, there are currently many standards such as DublinCore[3], IEEE LOM[8], etc. The existence of standards facilitates the management of the resources, enabling the interoperability among systems that use compatible standards.

Although at first sight these standards can be seen as an advantage, reality shows that in some cases they are the problem, as many existing standards are not compatible among themselves. The ADLNet[1] initiative was developed in order to solve these problems and to coordinate the effort of metadata standards and, in general, the use of IT in the educational context. It is important to note that not only is the existence of metadata standards necessary in order to reuse contents, but the data that the authors assign to each descriptor is very important as well. To this end, it is necessary to follow a traceable process from the creation of an educational resource to the creation of its metadata in order to establish a metadata structure that is consistent, relevant and interpretable. [2].

As with traditional education resources, LOs are stored in libraries, in this case digital libraries called repositories. A digital repository can be defined as a place

---

[1] Advanced Distributed Learning Network. http://www.adlnet.org/

where a digital resource can be stored, searched and retrieved. These systems must also support the import, export, identification and retrieval of content. [11]. The LOs are usually stored in a Learning Object Repository (LOR). JORUM project [10] states that an LOR is a set of LOs with detailed and external information (metadata) that is accessible from the Internet. In addition to housing the metadata, an LOR can also store the educational resource. In general terms, an LOR must implement the following task [¡**Error! No se encuentra el origen de la referencia.**]: Search/Find, Ask, Retrieve, Send, Store, Collect and Publish.

The deployment infrastructure can basically be either distributed or centralized. Taking into account that an LO is formed by a digital resource and its metadata, there are four kinds of possible infrastructures [6]: (i) centralized resources and centralized metadata, (ii) centralized resources and distributed metadata, (iii) distributed resources and centralized metadata and (iv) distributed resources and distributed metadata. Furthermore, three kinds of storage strategies can be distinguished [6]: (i) *File-based,* which uses files with predefined formats and an index-based management; (ii) *Database-based,* which uses any kind of database, and is the most extended method; and (iii) *Persistent objects-based,* where the LO are stored as serialized objects.

A controversial aspect is the interoperability among LORs. Firstly, it is necessary for a repository to use at least one standard in the stored metadata. And secondly, there must be an interface from where an external search agent (a client or another LOR) can access the stored information. This interface is currently implemented through high-level interoperability layers that allow access from outside the LOR. There are different standards or specifications that focuses on this interoperability layer:

- **OAI-MPH (*The Open Archives Initiative Protocol for Metadata harvesting*)[12].** This protocol provides a technology-independent framework for retrieving documents or resources, thus enabling interoperability among systems. The protocol is open and the repository is not limited to educational resources.
- **IMS DRI (*IMS Digital Repository Interoperability*) [9].** This is based on existing communication technologies and on previous specifications of the IMS consortium. This protocol is immature and is still in its initial stages of development.
- **SQI (*Simple Query Interface*) [4].** The kernel of SQI is formed by a set of abstract methods based on web services. These methods are not associated with any underlying technology. It is also is neutral in terms of the format of results as well as query language. This interfaces supports synchronous/asynchronous and stateful/stateless queries. The authentication is based on a session with the aim of isolating the harvesting of contents from the management tasks.

The use of an abstraction layer between LOR and the client system avoids the need to take the internal infrastructure of the repository into account, and allows clients to perform queries to many repositories in parallel. In other words, the clients can perform federated searches [1][5].

2.1 The Reality of LOR Interoperability

The state of the art shows a high heterogeneity in existing standards. Therefore, a study of LOR has been performed in order to analyze the real situation. The study includes the analysis of the following LORs: *Acknowledge, Agrega, Ariadne, AriadneNext, CGIAR, EducaNext, LACLO-FLOR, LORNET, MACE, Merlot, Nime, OER Commons* and *Edna Online*. It consists of performing 60 queries to each LOR through an SQI layer that the repositories provide. The query patterns are Unesco codes.

Firstly, the general characteristics of each LOR are analyzed; all of them use IEEE LOM as metadata schema and VSQL [15] as query language. Additionally, the majority of them are stateless (65%), and all of them have synchronous interfaces, but only 4 have the asynchronous interface.

Considering that the SQI specification does not force the implementation of all methods of the specification, the SQI methods that these repositories have implemented are checked. The results radically change the outlook because 6 of the 14 repositories do not work or are unavailable and they have to be removed from the scope of this study (Ariadne, AriadneNext, EducaNext, Nime and EdNa Online). MACE and LOCLO-FLOR produce an error in the authentication, although the process is done correctly. After this step, this test is reduced to only four repositories Acknowledge, Agrega, LORNET and Merlot. The latter three are perfectly valid and all SQI methods work perfectly; however the repository Acknowledge only implements the essential methods to perform queries.

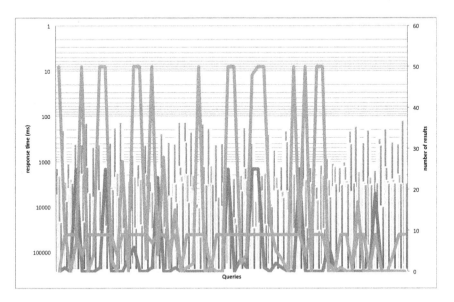

Fig. 1 Repositories performance

Finally, to summarize this study, Figure 1 shows the performance of each LOR. The lines are the number of results and the bars are the response times. The stateful repositories are slower than stateless repositories, with an average 19,1685 seconds compared to 3.053 seconds respectively, because they have to authenticate the client before performing the query. The average of the results is 9.56 LOs retrieved per query.

3   An Opportunity Focused on Cloud Computing

As it is possible to observe, the performance of the LOR is not appropriate. In order to deal with this problem, new LOR architectures have to be proposed and developed. This new generation of LOR must ensure the availability of resources and interoperability, permitting federated searches from external clients.

Lately, within the services in the context of Internet, Cloud Computing is emerging as key paradigm of the present century. According to NIST[2] [13], *Cloud computing is a model for enabling ubiquitous, convenient, on-demand network access to a shared pool of configurable computing resources (e.g., networks, servers, storage, applications, and services) that can be rapidly provisioned and released with minimal management effort or service provider interaction. This cloud model is composed of five essential characteristics, three service models, and four deployment models.* This definition includes three levels of computational services (Software, Platform and Infrastructure).

However, beyond the kinds of services, the key characteristic of this new paradigm is the quality of services. Cloud services are able to offer the same level of quality independently of instant demand. In practice, end users make use of Cloud services that are always available and unlimited.

Taking into account the weakness that has been demonstrated in this study with regard to the performance, availability and interoperability of existing LORs, this study proposes a new architecture based on Cloud Computing.

This new architecture will make use of the services that +Cloud platform [14] provides, such as storage and databases. This platform is based on the Cloud Computing paradigm. This platform allows offering services at the PaaS and SaaS levels. The IaaS layer is composed of a physical environment that allows the abstraction of resources into virtual machines.

The SaaS layer is composed of the management applications for the environment (virtual desktop, control of users, installed applications, etc.), and other more general third party applications that use the services from the PaaS layer. The components of this layer are:

- the *IdentityManager*, which is the module of +Cloud in charge of offering authentication services to clients and applications.
- the *File Storage Service (FSS)*, which provides an interface for a container of files, emulating a directory structure in which the files are stored with a set of metadata, thus facilitating retrieval, indexing, search, etc.

---

[2] NIST, National Institute of Standards and Technology (http://www.nist.gov/)

- the *Object Storage Service (OSS)*, which provides a simple and flexible sche-
maless data base service oriented towards documents.

3.1 Proposal: CLOR

This study proposes the development of a new platform called CLOR (Cloud-
based Learning Object Repository) based on +Cloud as its underlying architecture.
Figure 2 shows a diagram with the main components of this modern architecture.
The details of each component are presented as follows:

- **CLOR Management** is the kernel of the architecture. It is framed at the platform level within Cloud services. Its main task is to encapsulate the communication with the lower layers of the Cloud platform; at the same time, it is in
charge of providing a set of functionalities in terms of web services to the upper
layers, that is, to the end user interfaces.
  o FSS will be used to store the educational resources. FSS also encapsulates
    the traditional complexity of the file system storage; this component only
    has to call web services in order to retrieve or store resources. Furthermore,
    because of other FSS functionalities, such as file versions, metadata asso-
    ciated with each resource, etc., it will be possible to increase the power of
    the service. Finally, it should be noted that the elasticity of the FSS implies
    no limitation regarding storage capacity.
  o OSS will be used to store the metadata associated with each learning re-
    source. OSS makes use of a nonSQL database that permits storing the me-
    tadata en JSON format. The main advantage is that it permits storing any
    kind of metadata independent of its structure or schema, that is, its stan-
    dard. Furthermore, queries about the LO will be performed very quickly
    thanks to the underlying database.
  This component will be complemented with different interoperability layers,
  such as SQI or OAI-MPH, which will ensure the communication with other
  LORs and federated searches from external clients.
- **CLOR GUI** is the graphical user interface of the end users. The key architec-
tural characteristic is that it will be independent of the bottom layer (in other
words, the CLOR Management) and communication will be carried out through
web services. The management of the users will be delegated to the Cloud, specifically, to the identity manager.
  CLOR will present two independent interfaces:
  o *Storage CLOR* in charge of managing the repository (storage and creation of
    metadata).
  o *Search CLOR* will provide an interface to perform queries not only in this
    proposed repository, but also in the resources of other resources by means
    of the interoperability layers. This interface component will, in the future,
    also provide other functionalities such as a recommendation system, a key
    functionality that has recently emerged [1].

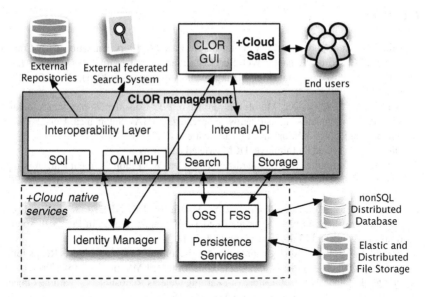

Fig. 2 CLOR architecture

4 Discussion, Advantages and Future Work

This study has presented an innovative architecture that constitutes an evolution over current storage system for educational resources. This new model, will enable the observed problems, which have been demonstrated in this study, to be solved:

- *High heterogeneity in terms of number and characteristics of existing standards.* The proposed model allows dealing with the heterogeneity of current and future standards since it is based on a non-relational database.
- *Low performance.* Cloud computing paradigm allows offering services with the same level of quality independently of its demand. The development of the LOR based on this paradigm will make it possible not only to offer an effective service effective, but to offer an unlimited storage capacity as well.
- *Interoperability among repositories.* The low linkage among components permits implementing many interoperability layers without needing to upgrade to other modules.
- *Complementary services.* This model will make it possible to include other functionalities in its own repository that until now were not possible, such as recommendation model, space of storage for each user in the cloud, a collaborative model for creating learning resources and metadata, etc.

Future work will be focused on finishing the development of the proposed repository and evaluating it not only at a technical level, but also at a functional level by using end users that are teachers and students.

References

1. Gil, A.B., Rodríguez, S., de la Prieta, F., De Paz, J.F.: Personalization on E-Content Retrieval Based on Semantic Web Services. International Journal of Computer Information Systems and Industrial Management Applications 5(2013), 243–251 (2010) ISSN 2150-7988
2. Berlanga, A.J., López, C., Morales, E., Peñalvo, F.J.: Consideraciones para reforzar el valor de los metadatos en los objetos de aprendizaje (OA). Salamanca. Universidad de Salamanca. Depto. deInformática y Automática (2005)
3. Dublin Core Metadata Initiative. DCMI Metadata Terms
4. European Committee for standardization – Cen Workshop Agregament. A simple Query Interface Specification for Learning Repositories. Ref. No.: CWA 15454:2005 E (2005)
5. De la Prieta, F., Gil, A.B., Rodríguez, S., Martín, B.: BRENHET2, A MAS to Facilitate the Reutilization of LOs through Federated Search. Trends in Practical Applications of Agents and Multiagent Systems, 177–184
6. Frango, I., Omar, N., Notargiacomo, P.: Architecture of Learning Objects Repositories. Learning Objects: standards, metadata, repositories & LMS, pp. 131–155 (2007)
7. Frango, I., Omar, N.: Architecture of Learning Objects Repositories. Learning Objects: standards, metadata, repositories & LMS, 131–155 (2007)
8. IEEE Learning Objet Metadata (LOM). Institute of Electrical and Electronics Engineers (2002), http://ltsc.ieee.org
9. IMS Digital Repositories Interoperability. Riley, K., McKell, M., y Mason, J. Core Functions Information Model. Version 1.0 Final Specification (2003)
10. JISC Online Repository for [learning and teaching] Materials (2004)
11. Joint Information System Committee. What is a Digital Repository? Digital Repositories. Helping Universities And Colleges (2010)
12. Lagoze, C., Van de Sompel, H., Nelson, M., Warner, S.: The Open Archives Initiative Protocol for Metadata Harvesting (2002)
13. Mell, P., Grance, T.: The NIST definition of Cloud Computing. In: NIST Special Publication 800-145 (September 2011)
14. Heras, S., De la Prieta, F., Julian, V., Rodríguez, S., Botti, V., Bajo, J., Corchado, J.M.: Agreetment technologies and their use in cloud computing environments. Progress in Artificial Intelligence 1(4) (2012)
15. Simon, B., Massart, D., Van Assche, F., Ternier, S., Duval, E.: Authentication and Session Management. Version 1.0. (2005)

Technological Platform to Facilitate the Labor Integration of People with Auditory Impairements

Amparo Jiménez, Amparo Casado, Elena García, Juan F. De Paz, and Javier Bajo

Abstract. This paper presents a technological platform aimed at obtaining an on-line workspace for exchanging digital contents in an easy, intuitive and accessible manner. The main objective of the platform is to provide facilities to inform, train and evaluate the competencies of disabled people, and more specifically those skills required to facilitate the labor integration of individuals with auditory disabilities. The platform also focuses on providing training processes that facilitate the incorporation of disabled people to labor environments. The platform presented in this paper has been tested in a real environment and the results obtained are promising.

Keywords: disabled people, auditory disability, competence, intelligent systems, learning and training processes.

1 Introduction

Education and training for disabled people has acquired a growing relevance during the last decade, especially for labor integration. Information and communication

Amparo Jiménez · Amparo Casado
Universidad Pontificia de Salamanca, Salamanca, Spain
e-mail: {ajimenezvi,acasadome}@upsa.es

Elena García · Juan F. De Paz
Departamento de Informática y Automática, Universidad de Salamanca
e-mail: {elegar,fcofds}@upma.es

Javier Bajo
Departamento de Inteligencia Artificial, Universidad Politécnica de Madrid
e-mail: javier.bajo@upm.es

J. Casillas et al. (Eds.): *Management Intelligent Systems,* AISC 220, pp. 107–117.
DOI: 10.1007/978-3-319-00569-0_14   © Springer International Publishing Switzerland 2013

technologies play a very important role in this evolution. Disabled people represent a considerable percentage of the current population and require special education.

During the past year, we have developed a research project linked to two different realities: professional training and proper professional performance with the special needs of some people with difficulties to access to employment. We selected as a target group the auditory impaired people and we focused in an specific job profile and job performance: Auxiliary Operations and General Administrative Services.

Subsequently we developed a technology-based training tool that allows the disabled people to effectively develop their professional performance, as well as to improve the professional training previous to the integration into the labor environment.

Thus, we propose a technological platform that focuses on obtaining on-line workspace for exchanging digital contents in an easy, intuitive and accessible manner. The main objective of the platform is to provide facilities to inform, train and evaluate the competencies of disabled people, and more specifically those skills required to facilitate the labor integration of individuals with auditory disabilities. This process may take place in the workplace or in the place of address via television, computer and mobile phone.

The rest of the paper is structured as follows: Section 2 presents the background of the problem taken into consideration. Section 3 presents the problem formalization. Section 4 describes the proposed technological platform. Finally, Section 5 presents the preliminary results and the conclusions obtained.

2 Background

First of all we must define the group taken into consideration in this study. As indicated by Calvo Prieto [1], the term auditory disabled refers to anyone who sees the word without sharpness enough or as indicated by the World Federation of the Deaf (Word Federation of the Deaf, EUD) person with hearing difficulty that can be alleviated with technical aids (FESORCV) [2].

We understand as those for people with a degree of disability (now disabled) greater than 33% by deafness or hearing limitations that encounter communication barriers (Spanish Law 27/2007, of 23 October, recognizing the Spanish sign languages and regulates the means of support for oral communication of the deaf, hearing impaired and deafblind, 2007) [3], or as a term currently used for the Confederation of Deaf People (CNSE) [4] or the Spanish Confederation of Deaf Families. (FIAPAS, 2004) [5].

To establish a proper design of professional guidance, we need to perform a detailed analysis of the specific characteristics of the position, profile and skills associated with their good work performance. Professional qualification is the "set of skills with significance in employment that can be acquired through training or other types of modular training, and through work experience" (Spanish Law 5/2002 on Qualifications and Vocational Training) [6].

From a formal point of view, the qualification is the set of professional competencies (knowledge, skills, abilities, motivations) that allow us to perform occupations and jobs with a valuable labor market impact and that can be acquired through training or work experience.

It implies, as noted by Levy-Leboyer [7] "a set of observable behaviors that are causally related to a good or excellent performance in a specific job and a specific organization" covering the full range of their knowledge and skills in personal , professional or academic, acquired in different ways and at all levels, from basic to top.

3 Problem Formalization

To formalize the problem we have relied on the National Catalogue of Professional Qualifications and professional qualifications, and we have selected the professional qualification Auxiliary Operations and Administrative Services for the Family General Administration and Management Professional with Level 1.

Our aim is not to find a professional qualification (which corresponds to the Initial Professional Qualification Programmes-PCPI-) but, based on the characteristics and requirements related to this position as identified in the Royal Decree 229/2008 of 1 February (BOE, No. 44 of February 20, 2008) [8], to identify some actions, strategies and more appropriate training resources, technologically updated and valid for the training and evaluation of the disabled individuals.

From our point of view, it is essential to follow the determination of the legal requirements and current proposals in the employment context. This allow us to train competent workers taking into account the parameters required in our socio-labor context, as well as the parameters shared by any worker (with or without disabilities) to develop such activities. We define, therefore, and employment and social integration strategy for people with different skills but that can afford with guarantees the demands of the position. Therefore, we respect the design of general competencies, skills units and professional achievements with performance criteria proposed in the Spanish National Catalogue of Professional Qualifications, as well as the different existing guidelines in Spain and those proposed by various international organizations.

Taking as starting point the document from the Spanish National Institute of Vocational Qualifications, we define an structure of the professional qualifications that will serve to design programs, resources, methodologies and educational interventions. In this sense, we have made a major effort to assign to each qualification a general competence. This competence includes the roles and functions of the position and defines the specific skills or competency units.

Described also the professional environment in which you can develop the skills, relevant productive sectors and occupations or jobs relevant to access it.

We have also described the professional labor environment in which the qualification will be evaluated, relevant productive sectors and occupations or jobs relevant to accessible obtaining the qualification.

Furthermore, in a complementary manner, we analyzed the professional achievements for each unit of competence along with their performance criteria.

We started with the following situation:

AUXILIARY OPERATIONS FOR ADMINISTRATIVE AND GENERAL SERVICES professional qualification.

General competence:

To distribute, reproduce and transmit the required information and documentation in the administrative and management task, internal and external, as well as to perform basic verification procedures on data and documents when senior technicians require it. These tasks are carried out in accordance with the existing instructions or procedures.

Competence units:

* To provide support for basic administrative operations.

* To transmit and receive operational information to external agents to the organization.

* To perform auxiliary operations for reproduction and archiving data on conventional computational support.

Professional field:

This individual operates as an employee in any company or private/public entity, mainly in offices or departments oriented to administrative or general services.

Productive Sectors:

It appears in all the productive sectors, as well as public administration. It is necessary to remark the high degree of inter-sectoriality.

Relevant occupations and possitions.

* Office Assistant.

* General services assistant.

* File assistant.

* Mail classifier and / or message.

* Ordinance.

* Information Assistant.

* Telephonist.

* Ticket clerk.

Fig. 1 Auxiliary Operations for Administrative and General Services professional qualification

However, we considered that, looking for a more specific training support, it is necessary to complete this information with the detailed description of the most common tasks that arise in professional performance. Thus, describing the specific tasks, we have established the type of support that this group of disabled people requires to carry out an effective performance of the assigned tasks. Finally, we have established the most appropriate training strategies. Thus, we have described the most common tasks related to the professional profile and professional qualification presented in the previous table. The following example illustrates our proposal:

Competence Unit: To provide support for basic administrative operations
Professional Development 1: To periodically register the Information updates of the organization, department, areas, personnel, according to the instructions previously received, with the aim of obtaining key Information to improve the existing services. Realizar un organigrama de la empresa o departamento

1. Make a list of phone and fax references of the various members of the company
2. Update the directory of people
3. To register the physical location of people and areas within the company.
4. To update the physical location of people and areas within the company.
5. Safe-keeping of keys.
6. Opening and closing the workplace and departments.
7. Bring documentation to other centers in the city (unions, Delegation, City Council, County Council, etc.)
8. Turn off and turn on the lights
9. Opening and closing windows
10. Open and lock any room.
11. To register the inputs and outputs of the employees.
12. To register a list for people who want to take the annual medical review.

4 Technological Platform

Based on the problem formalized in section 3 we obtained a technological platform. It is a software platform specifically designed to create intelligent environments [11] oriented to facilitate the labor integration of people with auditory disabilities. The main objective of the Ambient Intelligence is to achieve transparent and ubiquitous interaction of the user with the underlying technology [11]. In this paper we use Ambient Intelligence to design a software technology specialized on determining the professional qualification, and providing on-line tools focused on transmitting signed orders that are easily accessed via mobile devices. Basically, the proposed platform consists of a training tool via web, and a communication tool to send signed orders via mobile phone. In the following points we describe the main elements of the application.

– Order signing
Once the competences to evaluate were identified, and the related professional developments were defined, we proceeded to signing the actions and tasks that can be performed by the disabled person.

To make the signing we counted on the cooperation of the Federation of the Deaf of Castile and Leon (FAPSCyL) [12], who have participated in the signing process. The process followed consisted on recording a series of videos in which the sign interpreters transmit specific orders for each of the actions to be carried out by the disabled person. The recording was done in blocks, taking into account the professional developments taken into consideration.

Once the recording process finished, we proceeded to edit the videos obtained by separating each action individually and including subtitles in Spanish.

– **Web platform**

In this task we obtained the design and development of a web platform that allows us to transmit work orders to the auditory disabled person using sign language format. The orders are transmitted via the Internet, television or mobile devices. This is a simple Web page, based on Drupal CMS [13], which consists of the following sections:

Fig. 2 Main page of the platform for auditory disabled people

Figure 2 shows a web platform with differnt elements: a heather, menus for Home, Signed learnig, Learning evaluation and Contact:

- Home. It is the door to the web platform. It provides a user management section. A user authentication is required. Besides, this section describes the main objective of the platform and provides general instructions about its usability.
- Signed learning, provides instructions and exercises that are presented to the auditory disabled user bya means of videos. The videos show a signing interpreter transmiting orders.
- Learning evaluation. In this section a series of series are presented to the user. The surveys allow us to evaluate the user satisfaction degree and the effectiveness of the learning process.
- Contact. It provides the contact details for the platform.

The appearance of the platform is simple, trying to facilitate the accessibility and usability. The navigation through menus and contents is easy and intuitive. All the pages have been designed with the same structure, trying to facilitate a familiar environment and similar interaction patterns independently of the page or section in the platform. The next paragraphs describe the contents included in the platform:

Fig. 3 Introductory video

Figure 3 shows the content of the initial video in which the user can obtain a description of the operation of the application, as well as the use of videos containing the signed orders that correspond to the professional developments to perform in the workplace.

Once the user be in the learning section, the learning process is started, displaying the videos for the different blocks of accomplishments that can occur in the office environment:

Fig. 4 Learning section

– Mobile application

In this work we have designed and developed a mobile application for the platform that allows quick transmission of orders in the office workplace. The application includes voice recognition [16], so that a person at work may transmit voice instruction. These instructions will be recognized by the mobile device, which accesses a remote server and display the video corresponding to the order in sign language.

We revised the related work and the existing technologies to choose the best option for the mobile module. An analysis was made of all mobile platforms on the market to see which is more suited to our requirements. The module was developed for iOS [14], and can be installed on a device like iPad iPhone, as long as it has the same operating system version iOS 5 or above. This module uses an XML file [15] containing the structure of the data to be displayed. This XML file is stored in the cloud, and is parsed by our application. When the application starts, it parses the file and inserts into a table all the blocks, so that the user can choose one of them. Once the user clicks on a block, a screen containing an explicative video will be shown. The videos are also stored in the cloud. The advantages of using a cloud storage are that the content can be updated very easily and without jeopardizing the proper functioning of the application. Below, some screenshots for the developed application are presented, showing its operation.

Fig. 5 Main screen of the application

Figure 5 shows the main screen of the application that shows the user at the time of initiation. Once the user clicks on this screen, he can access the playback menu.

Fig. 6 Blocks scheme of the application

Figure 6 shows the screen with the touch playback options. Playback can be done in blocks or individually. Figure 6 shows a reproduction in blocks.

Fig. 7 Detail of a block

Figure 7 shows the playback screen, where the user has the option to view videos. The user can return to the playback of videos or play video.

Fig. 8 Example of signed vídeo

Figure 8 shows the playback screen, where the user is viewing a video. The application was initially developed for iPhone and iPad devices, but can be easily adapted to be executed on mobile devices.

5 Conclusions

The new model of labor relations established by the Spanish Royal Decree-Law 3/2012, of 10 February, on urgent measures for labor reform has among its objectives the promotion of inclusion in the labor market of more advantaged groups, including the people with disabilities. [16].

Our aim was to contribute to this goal by means of a technological platform specifically designed to facilitate labor insertion in office environment of people with auditory impairment. The proposed platform has a web interface and an interface for mobile devices, and is based on pre-recorded videos that contain instructions on actions to be performed by the disabled person in the office environment. The web interface was successfully tested in teaching through television, in collaboration with the company CSA and the results have been promising. The application was developed using the Drupal content management system and the iOS5 environment. We reviesed various content management and mobile platforms, leading to the conclusion that drupal and iOS5 were the most appropriated. However, it is necessary to indicate that the platform is easily adaptable to other technologies. Moreover, the mobile application was tested in an office environment. Users and FAPSCyL specialists have highlighted the utility and advantages of the application. A test was designed with 10 basic tasks performed by 3 disabled people before and after the platform presented in this paper was installed. The platform provided a new tool that contributed to increase the percentage of completed tasks up to 85%, when the initial percentage (without the platform) was 42%. The disabled users have remarked the ease of understanding of instructions they receive from their supervisors and ease of use of the system.

Acknowledgments. This work has been supported by the Ambientes Inteligentes con Tecnología Accesible para el Trabajo (AZTECA), CDTI. Proyecto de Cooperación Interempresas. IDI-20110345.

References

1. Calvo Prieto, J.C. (ed.): La sordera. Un enfoque socio-familiar. Amarú Ediciones, Salamanca (1999)
2. CNSE. Las personas sordas en España. Situación actual. Necesidades y demandas. Confederación Nacional de Sordos de España, Madrid (1996)
3. CNSE. Retos para el siglo XXI: Resoluciones del II Congreso de la Confederación de Sordos de España. Confederación Nacional de Sordos de España, Madrid (1998)
4. FESORCV. Minguet, A. (Coord.): asgos sociológicos y culturales de las personas sordas: una aproximación a la situación del colectivo de Personas Sordas en la Comunidad Valenciana. Federación de Personas Sordas de la Comunidad Valenciana, FESORD C.V., Valencia (2001)

5. FIAPAS. Jáudenes (Coord.): Manual Básico de Formación Especializada sobre Discapacidad Auditiva. Confederación Española de Padres y Amigos de los Sordos, Madrid (2004)
6. Levy-Leboner, C.: Gestión de competencias. Gestión 2000. Barcelona (1997)
7. Ley 27/2007, de 23 de octubre. Madrid: BOE nº 255, del 24-10-2007, pp. 43251–43259 (2007)
8. Ley 5/2002, de 19 de junio de las Cualificaciones y de la Formación Profesional. BOE del 20 de junio de, Madrid (2002)
9. Real Decreto 229/2008, de 1 de febrero, que recoge siete cualificaciones de la Familia profesional Administración y Gestión. BOE del 25 de febrero de, Madrid (2008)
10. Real Decreto-Ley 3/2012 de 10 de febrero, de medidas urgentes para la reforma del mercado laboral. BOE del 11 de febrero de, Madrid (2012)
11. Weiser, M.: The computer for the 21st century. Scientific American 265(3), 94–104 (1991)
12. Federación de Asociaciones de Personas Sordas de Castilla y León (FAPSCyL) (2012), http://www.fapscl.org/
13. Buytaert, Dries: Drupal (2012), http://drupal.org/
14. Napier, R., Kumar, M.: iOS 5 Programming Pushing the Limits. Wiley (2011) ISBN: 978-1119961321
15. Goldfarb, C.F., Prescod, P.: XML Handbook with CD-ROM. Prentice Hall PTR, Upper Saddle River (2001)
16. Reynolds, D.A.: An overview of automatic speaker recognition technology. In: 2002 IEEE International Conference on Acoustics, Speech, and Signal Processing (ICASSP), vol. 4, pp. 4072–4075 (2002)

Keystrokes and Clicks: Measuring Stress on E-Learning Students

Manuel Rodrigues, Sérgio Gonçalves, Davide Carneiro, Paulo Novais, and Florentino Fdez-Riverola

Abstract. In traditional learning, teachers can easily get an insight into how their students work and learn and how they interact in the classroom. However, in on-line learning, it is more difficult for teachers to see how individual students behave. With the enormous growing of e-learning platforms, as complementary or even primary tool to support learning in organizations, monitoring students' success factors becomes a crucial issue. In this paper we focus on the importance of stress in the learning process. Stress detection in an E-learning environment is an important and crucial factor to success. Estimating, in a non-invasive way, the students' levels of stress, and taking measures to deal with it, is then the goal of this paper. Moodle, by being one of the most used e-learning platforms is used to test the log tool referred in this work.

Keywords: E-learning, Behavioral Analysis, Stress, Moodle.

1 Introduction

When a student attends an electronic course, the interaction between student and teacher, without all its non-verbal interactions, is poorer. Thus the assessment of feelings and attitudes by the teacher becomes more difficult. In that sense, the use of technological tools for teaching, with the consequent teacher-student and student-student separation, may represent a risk as a significant amount of context

Manuel Rodrigues
Informatics Group, Secondary School Martins Sarmento, Guimarães, Portugal

Manuel Rodrigues · Florentino Fdez-Riverola
Informatics Department, University of Vigo, Ourense, Spain

Sérgio Gonçalves · Davide Carneiro · Paulo Novais
Informatics Department/ Computer Science and Technology Center,
University of Minho, Braga, Portugal

J. Casillas et al. (Eds.): *Management Intelligent Systems,* AISC 220, pp. 119–126.
DOI: 10.1007/978-3-319-00569-0_15   © Springer International Publishing Switzerland 2013

information is lost. Since students' effectiveness and success in learning is highly related to their mood while doing it, such issues should be taken into account when in an E-learning environment. In a traditional classroom, the teacher can detect and even forecast that some negative situation is about to occur and take measures accordingly to mitigate such situation. When in a virtual environment, such actions are impossible.

Stress, in particular, can play an important (usually negative) role in education [1-2]. In that sense, its analysis in an E-learning environment assumes greater importance. Using physiological sensors could be a solution for stress detection. However, the use of visible and invasive sensors induces itself a certain degree of stress. In this work, we extract information from keyboard strokes and mouse movement to generate important information about students' mood towards learning. We are developing a modular tool, able to estimate the level of stress of human users in a non-intrusive way. Our goal is to develop a dynamic stress estimation model that, while making use of this context information, will allow teachers to adapt strategies in search for increased success in learning.

2   Dynamic Student Assessment Module

As stated, stress has a significant influence in E-learning performance. To mitigate such problems, several research studies have been carried out. In [3,4] frameworks are proposed where the goal is to obtain an external module to be linked to the Moodle platform, enabling the detection of student's affective states and learning styles in order to really know each student and present contents accordingly. A similar affective module will be responsible for gathering all this information, and derive students' mood (e.g. states of mind or emotion, a particular inclination or disposition to learn something) in order to present relevant clues for a personalization and recommendation module, to be developed in future work. Figure 1 depicts the Dynamic Student Assessment Module (DSAM). Not detailed in this work is the Personalization and Recommendation Module that will be subject of future work. Furthermore, particular attention will be given to stress detection through keyboard and mouse.

The Dynamic Student Assessment Module has two sub-modules: explicit assumption and Dynamic Recognition (implicit assumption), whose function is to detect student's mood, maintaining that information (actual and past) in the mood database. This information will be used by another sub-module, the affective adaptive agent, to provide relevant information to the platform and to the mentioned personalization module. This allows actual students' mood information to be displayed in the Moodle platform, and to be used to personalize instruction according to the specific student, enabling Moodle to act differently with different students, and also to act differently to the same student, according to his/her past and present mood. Here, we refer to mood as the actual "willing" of the student to learn, which incorporates his/her affective state, learning style and level of stress.

Each student interacts with Moodle from his/her own real environment, when attending a course. This environment is equipped with sensors and devices that acquire different kind of information from the student in a non-intrusive way. While the student conscientiously interacts with the system and takes his/her decisions and actions, a parallel and transparent process takes place in which this information is used by the Dynamic Student Assessment Module. This module, upon converting the sensory information into useful data, allows for a contextualized analysis of the operational data of the students. This contextualized analysis is performed by the Dynamic Student Assessment Module. Then, the student's profile is updated with new data, and the teacher responsible for that course receives feedback from this module.

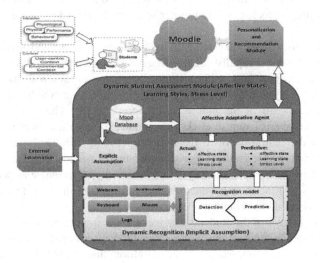

Fig. 1 Dynamic Student Assessment Module

To accomplish this, various research works of the research team are being integrated, touching areas such as facial recognition, behavioral analysis and noninvasive stress assessment. Two newly developed modules are explained next, with particular emphasis on mouse and keyboard logs.

2.1 Explicit and Implicit Mood Assumption

One of the easiest ways of knowing a student's mood is by making explicit questions. Surprisingly, this may not be the most accurate way: not always the answers obtained reveal the accurate state of the student. However, we can still use questionnaires as a way of gathering some useful information. An explicit mood assumption agent could periodically pose some questions, preferably in a visual way, for the student to upgrade his/her mood to the system. This configures the

Explicit Mood Assumption module. Several other research works have been carried out to detect student's mood explicitly [5].

A more interesting approach would be to infer such information. This is done under the Implicit Mood Assumption module. The aim of this sub-module is to monitor the interactions between the student and the system in order to infer the students' mood without being intrusive, that is, without the student being aware of the analysis being performed. Four key aspects are considered, although only two are in the scope of this this work: facial analysis, mouse analysis, keyboard analysis, and log analysis. As web cams tend to be standard equipment in computers nowadays, the goal is to use them to help infer emotions from the user. Mouse movements can also help predict the state of mind of the user, as well as keyboard usage patterns. Finally, analyzing the past interactions of the student through the logs of Moodle makes it possible to infer some of the information we are looking for.

The way a user types may indicate his/her state of mind and level of stress. Pressing keys hard and rapidly could indicate an altered state such as anger, while taking too much time may mean sadness. The same occurs with mouse movements. A similar system that monitors users' behavior from standard input devices, like the keyboard or the mouse, is proposed by [6]. The features analyzed include: the number of mouse clicks per minute, the average duration of mouse clicks (from the button-down to the button-up event), the maximum, minimum and average mouse speeds, the keystroke rate (strokes per second), the average duration of a keystroke (from the key-down to the key-up event) and performance measurements. [7] includes keyboard stroke information in order to improve the accuracy of visual-facial emotion recognition.

The level of stress of the students assumes greater importance due to its correlation to success. The focus of this work is thus on devices capable of acquiring data related to stress. The following sources of information (from now on designated sensors) acquired from the respective devices are:

- **Click accuracy** – a comparison between clicks in active controls versus clicks in passive areas (e.g. without controls, empty areas) in which there is no sense in clicking. This information is acquired from the mouse.
- **Click duration** – this represents the time span between the beginning and the end of the click event. This data is acquired from the mouse.
- **Amount of movement** – the amount of movement represents how and how much the student is moving inside the environment. An estimation of the amount of movement from the video camera is built. The image processing stack uses the principles established by [8] and uses image difference techniques to calculate the amount of movement between two consecutive frames [9] .
- **Mouse movement** – the amount of mouse movement represents the pattern in which the student moves the mouse (e.g. low amplitude quick movements of the mouse may indicate a high level of stress). This data are acquired from the mouse.

- **Mouse clicks** – the amount of mouse clicks and its frequency is useful for building an estimation of how much the student is moving around the screen and where he/she clicks. This data is acquired from the mouse.
- **Keyboard strokes** – frequency and intensity of the use of the keyboard. Frequent backspaces may indicate errors, high keyboard stroke may suggest experienced user (student) as opposed to low keyboard strokes. Stroke intensity (if keyboards allow it) may also be considered. This data is acquired from the keyboard.

Using an E-learning platform requires computers and peripherals for interaction. The most common peripherals nowadays are the keyboard and the mouse. Commonly, these interaction instruments are present in classrooms that are used to work with E-learning platforms and also in our homes and offices. Hence the increased advantage of their use for estimating stress.

Concerning the keyboard, there are currently many consolidated studies that point out the accuracy of keystroke analysis, allowing even users' recognition [10]. An area known as keystrokes dynamics aims at the recognition of users through the password provided as well as the correct rhythm of character input [11]. Thus, the changing of keystroke rhythm is a crucial factor to be considered in stress analysis [12].

The cases presented demonstrate how well the process of using keystroke log software packages is regimented. In a normal environment it is possible, using this type of approach in a transparent way, to collect data for later analysis. The user does not feel any intrusion into their actions using the e-learning environment.

The second peripheral that is used very often is the mouse. Its use is already taken for granted in interaction with learning platforms. Hence, the data analysis resulting from the use of this device is of utmost importance. There are already documented experiences of using mouse tracking software in areas of psychological analysis [13]. The coordinates (log coordinates of user interaction), the frequency of movements (log each time interaction) shows the state of the user, whether it will be stressed or normal.

Another area in which mouse tracking is fundamental and a key factor in decision-making is web browsing [14]. It is possible to log the cursor path that users create while browsing web pages. Presently it is also possible to measure the pressure exerted in a device such as a mouse [15]. This gives the possibility to detect actions of different pressure, of users in different psychological states. Mouse tracking should thus be, along with the use of the keyboard, one of the main points of data analysis for the determination of stress. It becomes clear that the possibilities for data analysis are feasible and reliable.

At this point one of the major difficulties and failures that Moodle presents is its log tool that complicates the analysis of actions and the knowledge of when that action took place. It is not possible to get, with the required certainty, the analysis of users' activity with the main focus on the type of interface used and the type of movement performed with this tool. In addition to this factor, there is the absence of date/time registration, so that the frequency of activity analysis could be determined. Thus, a log tool was developed in C# language. This is a simple but powerful application which enables to acquire data of users' actions and register them in a log file.

The record of the actions taken with the keyboard follows the logic of recording the type of action with key (Key Up or Key Down) and adding to the information the moment of key usage, in milliseconds.

Concerning the mouse, the registration of actions considers the movement, the clicks, the scrolls and, of course, the time which each of these events took place and their coordinates. This application, that is running in background, creates a log describing the interaction of the student with the Moodle platform.

3 Data Analysis

We used ten programming students, as our test group, in order to validate the possibility of detecting users' stress through the analysis of mouse and keyboard usage. The log tool mentioned earlier was used to assess the activity of this group when using a computer. The tested students were not aware of the existence of such tool running on the background of their computers. First, an assignment was given, with no time restrictions and with no influence in terms of difficulty to the student. The correspondent data was collected and analysed. Secondly, at a later moment, the same group was given an assignment, very similar to the first one, but with many constraints introduced. There was a time limit, they were told that the resulting work would have a major importance on their classification, and this group was intentionally submitted to stressing and upsetting factors. The results obtained are clarified in the following figures and table.

This allowed to derive knowledge from keyboard and mouse data. Concerning mouse movements, we found that in the first situation (where a student accomplished a task with no restrictions whatsoever and with no concerns of grade) there were considerably less mouse movements and keyboard usage than when the student had the mentioned restrictions (in terms of work volume, difficulty, influence on final grade), as stated in Table 1. A more detailed analysis of the gathered information while prosecuting the proposed activities allows the evaluation of the kind of usage of the two devices (mouse and keyboard).

Table 1 Type of movements during tasks

Type of movement	Stress less	Stressed
KD	9	30
KU	9	30
MD	31	50
MOV	2493	5427
MU	31	50
MW	43	208

KD: key down; **KU**: key up; **MD**: mouse down; **UM**: mouse up; **MW**: mouse wheel ; **Mov**: mouse movement

It is concluded from the mentioned data analysis that, when stressed, the number of mouse usage and keyboard pressing is substantially greater than that of a calm student. Mouse clicks and scroll usage shows a greater hesitation and jitters by the user when accomplishing the proposed task.

In keyboard usage, it is also easy to understand that in a stressed situation, the student uses this device more intensively, with the backspace key having a high frequency of use (Figures 2 and 3).

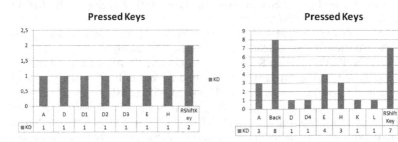

Fig. 2 Pressed Keys without stress   **Fig. 3** Pressed Keys in stressed situation

4 Conclusions

It is common sense that stress significantly influences learning capacities, thus learning success. This is particularly relevant when engaging an on-line course, using an E-learning platform, as several research lines indicate. To cope with this factor (together with others), when in an E-learning environment, a Dynamic Student Assessment Module was proposed. This DSAM proposes the use of several usual equipments as sensors, without the user being aware of them. In this paper, particular attention is given to keyboard and mouse, with the development of a log tool to monitor keyboard and mouse usage transparently. A group of programming students was then used to evaluate the hypothesis of stress detection trough mouse and keyboard activity. From the collected data and from posterior analyses it becomes clear that it is highly feasible to detect stress by this method. Significant work is still needed however, and a DSAM as the one proposed in this paper is being developed, to enhance E-learning students' success.

Acknowledgments. This work is funded by National Funds through the FCT - Fundação para a Ciência e a Tecnologia (Portuguese Foundation for Science and Technology) within projects PEst-OE/EEI/UI0752/2011 and PTDC/EEI-SII/1386/2012. The work of Davide Carneiro is also supported by a doctoral grant by FCT (SFRH/BD/64890/2009).

References

1. Carneiro, D., Novais, P., Neves, J.: Toward seamless environments for dispute prevention and resolution. In: Novais, P., Preuveneers, D., Corchado, J.M. (eds.) ISAmI 2011. AISC, vol. 92, pp. 25–32. Springer, Heidelberg (2011)

2. Palmer, S., Cooper, C., Thomas, K.: Creating a Balance. Managing Stress British Library, London (2003)
3. Rodrigues, M., Fdez-Riverola, F., Novais, P.: Moodle and Affective Computing - Knowing Who´s on the Other Side. In: 10th European Conference on E-learning, ECEL-2011, November 10-11, pp. 678–685. University of Brighton, Brighton (2011) ISBN: 978-1-908272-22-5
4. Fdez-Sampayo, C., Reboiro, M., Glez-Peña, D., Fdez-Riverola, F.: Sistema de seguimiento de actividades en moodle para la evaluación comparativa del ratio de participación alumno/clase. In: Conferencia Ibero-Americana WWW/Internet 2009, CIAWI 2009, Madrid, Spain, October 21 (2009)
5. Broekens, J., Jonker, C.M., Meyer, J.J.C.: Affective negotiation support systems. Journal of Ambient Intelligence and Smart Environments 2(2) (2010)
6. Zimmermann, P., Guttormsen, S., Danuser, B., Gomez, P.: Affective Computing-A Rationale For Measuring Mood With Mouse And Keyboard. International Journal of Occupational Safety and Ergonomics: JOSE 9(4), 539–551 (2003)
7. Tsihrintzis, G.A., Virvou, M., Alepis, E., Stathopoulou, I.-O.: Towards Improving Visual-Facial Emotion Recognition through Use of Complementary Keyboard-Stroke Pattern Information. Itng. In: Fifth International Conference on Information Technology: New Generation, pp. 32–37 (2008)
8. Castillo, J.C., Rivas-Casado, A., Fernández-Caballero, A., López, M.T., Martínez-Tomás, R.: A multisensory monitoring and interpretation framework based on the model-view-controller paradigm. In: Proceedings of the 4th International Workshop on the Interplay between Natural and Artificial Computation, vol. 1, pp. 441–450 (2011)
9. Fernández-Caballero, A., Castillo, J.C., Martínez-Cantos, J., Martínez-Tomás, R.: Optical flow or image subtraction in human detection from infrared camera on mobile robot. Robotics and Autonomous Systems 58(12), 1273–1281 (2010)
10. Dowland, P., Furnell, S.: A Long-Term Trial of Keystroke Profiling Using Digraph, Trigraph and Keyword Latencies. In: Deswarte, Y., Cuppens, F., Jajodia, S., Wang, L. (eds.) Security and Protection in Information Processing Systems, vol. 147, pp. 275–289. Springer, US (2004), doi:10.1007/1-4020-8143-X_18
11. Monrose, F., Rubin, A.: Authentication via keystroke dynamics. In: Proceedings of the 4th ACM Conference on Computer and Communications Security, pp. 48–56. ACM, New York (1997), doi:10.1145/266420.266434
12. Alves, F., Pagano, A., Da Silva, I.: A new window on translators' cognitive activity: methodological issues in the combined use of eye tracking, key logging and retrospective protocols. Copenhagen Studies in Language (38), 267–291 (2010), http://cat.inist.fr/?aModele=afficheN&cpsidt=22433271 (retrieved)
13. Freeman, J., Ambady, N.: MouseTracker: Software for studying real-time mental processing using a computer mouse-tracking method. Behavior Research Methods 42(1), 226–241 (2010), doi:10.3758/BRM.42.1.226
14. Arroyo, E., Selker, T., Wei, W.: Usability tool for analysis of web designs using mouse tracks. In: CHI 2006 Extended Abstracts on Human Factors in Computing Systems, pp. 484–489. ACM, New York (2006), doi:10.1145/1125451.1125557
15. Akamatsu, M., MacKenzie, I.S.: Changes in applied force to a touchpad during pointing tasks. International Journal of Industrial Ergonomics 29(3), 171–182 (2002), doi:10.1016/S0169-8141(01)00063-4

Integral Multi-agent Model Recommendation of Learning Objects, for Students and Teachers

Paula Rodríguez, Néstor Duque, and Sara Rodríguez

Abstract. Currently, there has been progress in building models for search and retrieval of learning objects (LO) stored in heterogeneous repositories. Likewise, research has increased the recycling of educational materials. This paper proposes the integration of two multi-agent models focused on delivering, specific LO adapted to a student's profile; and delivering LO to teachers in order to assist them in creating courses. The objective is to have an integral multi-agent model that meets the needs of students and teachers, and in this way improve the teaching learning process.

Keywords: Artificial Intelligence in Education, Multi-Agent Systems, Learning Objects, Repositories, Recommendation Systems, e-learning, Virtual organizations, Case Based Reasoning CBR.

1 Introduction

The value of information as a learning resource has created the need to share and reuse it without great cost, this added to the development of specifications and standards to solve the problem of incompatibility between different platforms has fueled the emergence of the Learning Object (LO) concept. A LO is a minimum unit of content used to teach something reusable on different platforms. The LO is distinguished from traditional educational resources because they are immediately available in a Web-based repository, to access them through metadata. With the intent of maximizing the number of LO that a user can use in order to support the teaching and learning process, centralized digital repositories are joined in federa-

Paula Rodríguez · Néstor Duque
Universidad Nacional de Colombia
e-mail: {parodriguezma,ndduqueme}@unal.edu.co

Sara Rodríguez
Universidad de Salamanca
e-mail: srg@usal.es

J. Casillas et al. (Eds.): *Management Intelligent Systems,* AISC 220, pp. 127–134.
DOI: 10.1007/978-3-319-00569-0_16   © Springer International Publishing Switzerland 2013

tions of repositories to share and access to other resources [1]. LO must be tagged with metadata, so they can be located and used for educational purposes in Web-based environments [2]. The use of metadata reinforces its utility because of the recovery, localization, exchanges and thus reuse is facilitated. Its importance lies in the fact that through the metadata the initial approach towards the resource can be made and the characteristics can quickly be learned. Recommendation systems are widely used online to support users in finding relevant information [3]. Having a user profile allows the identification of needs and preferences of the student who is doing a search for LO, in order to find the most relevant information or to give recommendations related to which LOs could support the learning process. Intelligent agents are entities that have sufficient autonomy and intelligence to handle specific tasks with little or no human supervision [4]. They are being used almost the same way as traditional systems, making it a good choice for solving problems where one needs autonomous systems to work individually and cooperate with each other to achieve a common goal [5].

The purpose of this paper is to integrate two approaches based on multi-agent systems, that work together to support the teaching and learning processes. The first of which pretend to provide teachers with a tool to recover educational resources to help organize courses [6]; and the second seeks is recommend the most pertinent LO according to the student profile [7]. With the aim of addressing the obstacles that appeared, we propose a model based on multi-agent systems based on virtual organizations that combine the advantages of the two approaches in which we have worked. The rest of the paper is organized as follows: Section 2 outlines main concepts involved in this research; Section 3 describes some works related to the proposed model. Section 4 introduces the multi-agent model proposal. Then, section 5 shows a case study where a course is associated with a teacher, and a student going to the course. LO is recommended from the repository according to the two profiles, a recommendation for a teacher to a student in particular, in order to validate the integration of the two approaches. Finally, conclusions and future work are presented in Section 6.

2 Basic Concepts

2.1 Learning Objects, Repositories and Federations

Wiley [8] defined LOs as the elements of a new type of instruction based on the object-oriented paradigm (OOP), typical of computer science. The IEEE [9] defines a learning object LO as: *any entity, digital or not with instructional design features, capable of being used, reused or referenced during computer-assisted learning.* The LO are contained in repositories (LOR) that are specialized digital libraries, they host many types of educational resources and their metadata, which are used in various e-learning environments. There are local repositories defined as digital libraries those specific to an organization that contain further LO own remote repositories that are accessed over a network. The project JORUM [10], proposes that a LOR is a collection of LO that has information (metadata) which is detailed and accessible via Internet. Besides housing the LO, the LOR can store the locations of those objects stored elsewhere, both online and in local locations. A federation provides a unified representation of these repositories [11], and serves to facilitate uniform administration of applications to discover and access content in the LO group LOR.

2.2 Recommender Systems

Recommendation Systems are aimed to providing users with search results that match their needs, making predictions of their preferences and delivering those items that could come closer than expected [12],[13]. In the context of LOs these systems seek to make recommendations according to the student's characteristics and their learning needs. Furthermore, in order to improve recommendations, recommender systems must perform feedback processes and implement mechanisms that enable them to obtain a large amount of information about users and how they use the LO [1],[14]. On the other hand, to help teachers create their virtual courses, recommender systems should help teachers in the design and creation of courses, suitable for delivering LO educational resources from reuse and sequencing learning activities [15].

The student profile stores information that can be used to obtain search results according to specificity. Managing a user profile aids the student or a teacher in the LO selection according to their preferences and personal characteristics [5].

2.3 Collaborative Filtering and Case Based Reasoning (CBR)

The CBR, is a kind of system that solves current problems, using the stored experience that corresponds to past problems. A basic recommendation problem is when a user wants to buy a product; the system recommends products that are related to the first search. If the user accepts the recommendations, this problem is stored. To use a CBR as a recommendation system it is necessary to add 2 to 4 stages of a classic CBR (Retrieve, Reuse, and Retain Check). Moreover, for the definition of this case, there is no predefined model, since it often depends on the problem, but usually includes information about the object, the user, the problem and the solution itself [16].

2.4 Multi-agent System

The agents are entities that have the autonomy to perform tasks and achieving their objectives without human supervision. This paradigm presents a new form of analysis, design and implementation of complex software systems and has been used for the development of recommendation systems [5].

The desirable characteristics of the agents are as follows [17]: autonomy, reactivity, pro-activity, cooperation and coordination, deliberation, distribution of tasks, adaptation, and parallelism.

2.5 *Virtual Organizations of Agents and Architecture*

The Virtual Organization (VO) is defined as a set of individuals and institutions, which need to coordinate their resources and services within institutional limits [18]. The VO can be considered as open systems formed by grouping and collaboration of heterogeneous entities where there is a clear separation between structure and functionality.

3 Related Works

Huang et al. present a framework for the creation of standardized courses, including an e-learning portal LMS, an authoring tool for editing educational resources, a navigation training system, which collects information from the user's learning history and gives course suggestions [19]. However, the two recommendations being executed in this work are not performed. Campos et. al. present a multi-agent model to search heterogeneous repositories with semantic features and information users. Besides, a LO classifier is presented which shows user information, statistics and evaluations of objects [20]. In the work of Li et. al. 2012, a ubiquitous learning system is proposed which is based on the student's learning records. This study uses personalized instruction and a context-sensitive method supported by a ubiquitous learning log [21], which only makes recommendations for students.

Casali 2011, presents a recommendation system based on intelligent agents, which aims to return a ranked list of the most suitable LO according to a user profile. The search is performed in the repository Ariadne [5]. The main limitation of this investigation is that despite some student's characteristics are considered in the user profile the user learning styles were not taken into account.

Rodríguez et al. 2012 propose a model for search, retrieval, recommendation and evaluation of LO from repository federation, called BROA, delivering of LO adapted to the student's profile, learning style and preferences, SMA is a student-focused recommendation [7].

De la Prieta et al. 2011, present BRENHET2 search of educational resources in heterogeneous, multi-agent systems to facilitate reuse of LO through federated search that incorporates a phase of cataloging and filtering results. Federated search is based on the application of virtual organizations SMA [6].

The last two models covering different approaches and looks promising them together to have comprehensive system.

4 Proposed Model

We propose a multi-agent system for search, retrieval, and recommendation of LO, for students and teachers. The LO results of the search, for students are recommended to the profile and learning style. For teachers the system facilitates the composition of training courses from cataloging LO. The search is performed in local and remote repositories, or federations of repositories, accessible via web and descriptive metadata of these objects. It keeps the MAS approach in order to exploit their advantages, among which are: parallelism, the ability of deliberation, Cooperation, Coordination and Distribution. In addition to the federated search process, the MAS has rules and social organizations due to the high heterogeneity of this context. Figure 1 shows the system architecture.

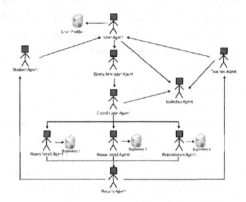

Fig. 1 System Architecture

To make the student's recommendation, using the student profile that consists of user information and preferences, and learning style; LO is recovered from the LOR and the search is performed by metadata title, description and keywords, to find the similarity with the preferences also a match is made between learning style and metadata: learning Resource Type, Interactivity Level, Intended End User Role, Definition, Description and LanguageSystem Architecture.

4.1 Agent Description

User Agent: Represents the human user into the system. It has the functions of starting the search process, validating results and also has access to statistics. It manages the user's profile allowing the creation and modification of the characteristics and preferences according to the profile. **Query Manager Agent:** Is the responsible of monitoring the entire search process. It has the functions to retrieve the query, start the search process, cataloging request and retrieve results from the query statistics. **Coordinator Agent:** This agent redirects the user queries to the search repositories. It has the specific control over the process specific federated search. Also, it should monitor the correct operation of these agents to maximize performance. **Repository Agent:** Repository agents are responsible of making the actual searches in the repositories. It implements the possible middleware code layers. There will be as many LOR agents, as repositories, which are directed toward the query. **Results Agent:** Receives the LO from each LOR agent. Automatically extracts metadata schema information and removes invalid LO, and finally calculates a reuse statistic. **Teacher Agent:** Is responsible for developing the ranking of results. It implements a CBR to develop a ranking of the LO will best suit the needs of the teacher on the basis of information previously obtained. **Student Agent:** Recommendations to students are made from the LO after user login query. This recommendation is based on the student's learning style. **Statistics Agent:** This agent is responsible of collecting statistical data from other agents (ROA, evaluation of results and LO). It communicates with the local LOR agent to store these evaluations; the relationship remains LO - evaluation - user.

5   Experiments and Results

To evaluate the proposal was made searches with MAS and crossed the recommendations made by the teacher agent and the student agent that would be valid for an integrated system. Multiple searches were performed with the two models, one of which is shown in Figure 2, the query string was cardiology, and Figure 2 shows the recommended LO system of teachers and students; on the right it shows the LO recommended by the proposed system.

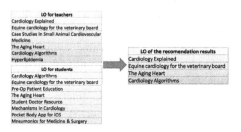

Fig. 2 Result LO list of the three recommendation system

Several tests were performed at 2 systems; the teachers and students with the same query string, as was done with the system proposed in this paper, the amount of recovered LO is shown in the Figure 3. With the proposed system are selected LO that are relevant to a course of a teacher where one student attends.

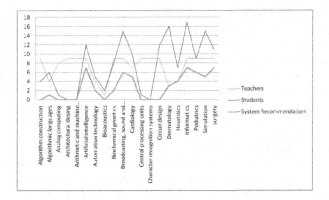

Fig. 3 Result of three recommendation system

6 Conclusions and Future Work

This first approach of integrating the two approximations (students and teachers) is promising have multi-agent system for recommending LO for students and teachers, operating together, recognize the user's needs and making a tailored recommendation, particularly to support the teaching-learning process. The system facilitates the composition of large training courses for teachers, by composing smaller elements. In addition, the system supports the learning of students based on their profile. The proposed MAS improves what already exists in the review found no recommendation systems that perform the two processes, helping teachers to develop new teaching materials and educational resources recommendation to students.

Currently it has developed a prototype for validation; as future work is to extend both theoretically and practically the MAS, using the advantages of both models: use of user profiles and collaborative filtering with CBR. Also defines the thresholds to determine the qualifications for making the most effective recommendation according to the two integrated approaches.

References

1. Li, J.Z.: Quality, Evaluation and Recommendation for Learning Object. In: International Conference on Educational and Information Technology, no. Iceit, pp. 533–537 (2010)
2. Gil, B., García, F.: Un Sistema Multiagente de Recuperación de Objetos de Aprendizaje con Atributos de Contexto. In: ZOCO 2007/CAEPIA (2007)
3. Niemann, K., Scheffel, M., Friedrich, M., Kirschenmann, U., Schmitz, H., Wolpers, M.: Usage-based Object Similarity. Journal of Universal Computer Science 16(16), 2272–2290 (2010)
4. Wooldridge, M., Jennings, N.R., Kinny, D.: A methodology for agent-oriented analysis and design. In: Proceedings of the Third Annual Conference on Autonomous Agents, AGENTS 1999, vol. 27, pp. 69–76 (1999)

5. Casali, Gerling, V., Deco, C., Bender, C.: Sistema inteligente para la recomendación de objetos de aprendizaje. Revista Generación Digital 9(1), 88–95 (2011)
6. De la Prieta, F., Gil, A., Rodríguez, S., Martín, B.: BRENHET2, A MAS to Facilitate the Reutilization of LOs through Federated Search. In: Corchado, J.M., Pérez, J.B., Hallenborg, K., Golinska, P., Corchuelo, R. (eds.) Trends in Practical Applications of Agents and Multiagent Systems. AISC, vol. 90, pp. 177–184. Springer, Heidelberg (2011)
7. Rodríguez, P., Tabares, V., Duque, N., Ovalle, D., Vicari, R.M.: Multi-agent Model for Searching, Recovering, Recommendation and Evaluation of Learning Objects from Repository Federations. In: Pavón, J., Duque-Méndez, N.D., Fuentes-Fernández, R. (eds.) IBERAMIA 2012. LNCS, vol. 7637, pp. 631–640. Springer, Heidelberg (2012)
8. Wiley, D.A.: Connecting learning objects to instructional design theory: A definition, a metaphor, and a taxonomy. Learning Technology 2830(435), 1–35
9. Learning Technology Standards Committee, IEEE Standard for Learning Object Metadata. Institute of Electrical and Electronics Engineers, New York (2002)
10. JORUM, JISC Online Repository for [learning and teaching] Materials (2004), http://resources.jorum.ac.uk/xmlui
11. van de Sompel, H., Chute, R.: The aDORe federation architecture: digital repositories at scale. International Journal 9, 83–100 (2008)
12. Chesani, F.: Recommmendation Systems. Corso di laurea in Ingegneria Informatica, 1–32 (2002)
13. Mizhquero, K.: Análisis , Diseño e Implementación de un Sistema Adaptivo de Recomendación de Información Basado en Mashups. Revista Tecnológica ESPOL (2009)
14. Sanjuán, O., Torres, E., Castán, H., Gonzalez, R., Pelayo, C., Rodriguez, L.: Viabilidad de la aplicación de Sistemas de Recomendación a entornos de e-learning. Universidad de Oviedo, España (2009)
15. Verbert, K., Ochoa, X., Derntl, M., Wolpers, M., Pardo, A., Duval, E.: Semi-automatic assembly of learning resources. Computers & Education 59(4), 1257–1272 (2012)
16. De, F., Pintado, P., Corchado, M.: Recuperación y Catalogación de Recursos Educativos mediante Organizaciones Virtuales y Filtrado Colaborativo 1 Introducción y objetivos (2011)
17. Jennings, N.R.: On agent-based software engineering. Artificial Intelligence 117(2), 277–296 (2000)
18. Boella, G., Hulstijn, J., Der Torre, V.: Virtual organizations as normative multiagent systems. In: HICSS. IEEE Computer Society (2005)
19. Huang, Y.-M., Chen, J.-N., Huang, T.-C., Jeng, Y.-L., Kuo, Y.-H.: Standardized course generation process using Dynamic Fuzzy Petri Nets. Expert Systems with Applications 34(1), 72–86 (2008)
20. Campos, R., Comarella, R., Azambuja, R.: Model of Recommendation System for for Indexing and Retrieving the Learning Object based on Multiagent System. In: MASLE 2012 (2011)
21. Li, M., Ogata, H., Hou, B., Uosaki, N., Yano, Y.: Personalization in Context-aware Ubiquitous Learning-Log System. In: 2012 IEEE Seventh International Conference on Wireless, Mobile and Ubiquitous Technology in Education, pp. 41–48 (March 2012)

TANGO:H: Creating Active Educational Games for Hospitalized Children

Carina Soledad González, Pedro Toledo, Miguel Padrón,
Elena Santos, and Mariana Cairos

Abstract. This paper presents an interactive platform called TANGO:H (Tangible Goals: Health) for the creation of social active play with gestural interaction. This platform has been specially designed and adapted for its recreational, educational and rehabilitation use in children at hospital. The platform has a game editor called TANGO:H Designer, which allows the creation of physical and cognitive exercises in single and multiplayer, so that children can play sequentially or simultaneously both competitively and collaboratively. Similarly, the design of the platform was taken into account various criteria of playability and gamification to maximize the user experience with the game. This article presents the requirements that have been followed in the design of Tango-H, and the developed solution.

Keywords: Exergames, Serious Games, Rehabilitation, Gamification, Kinect, OpenNI.

1 Introduction

The hospital classrooms (AH) provide educational services to students hospitalized during the period of compulsory education. The types of disease affecting these students are varied, but stand out the oncological, orthopaedic, respiratory, diabetes and surgery. The time of admission to hospital may be of short duration (up to 5 days), medium duration (6-20 days) and long duration (over 21 days) and

Carina Soledad González · Pedro Toledo · Elena Santos · Mariana Cairos
Department of Engineering Systems and Automation and Architecture and Computer Technology. University of La Laguna
e-mail: {carina,pedro,elena,mariana}@isaatc.ull.es

Miguel Padrón
Institute of Renewable Technology and Energy(ITER)
e-mail: mpadron@iter.es

J. Casillas et al. (Eds.): *Management Intelligent Systems,* AISC 220, pp. 135–142.
DOI: 10.1007/978-3-319-00569-0_17   © Springer International Publishing Switzerland 2013

may affect the process of socialization and formation of the child. An average of 210 children a year is treated in the seven Canary hospital classrooms. Collaboration and discussion with clinicians (Paediatrics, Nursing and Physiotherapy) on the needs of this hospitalized population concluded that it was feasible to develop playful tools based on ICT and a scenario to boost and strengthen the physical and motor activity of hospitalized children and adolescents [4]. Note that this is a facet of their development so far unattended, but there are several researches tools for the rehabilitation purposes [1][1]. These reasons have motivated the creation of the platform TANGO:H (Tangible Goals: Health), which in its design and development has two main objectives:

- Design and develop an open platform accessible and highly configurable to enable the creation, customization and adaptation of exercises and activities according to the specific characteristics of each user and user group.
- Design a social gaming platform, educational and of rehabilitation that follows the principles of game play and gamification to maximize motivation and customer satisfaction in its execution.

Below we will analyze the requirements that we have followed in the design and development of the platform, then will describe the principles of game play and gamification and how we have applied it in our application.

2 TANGO:H

TANGO:H is an application aimed to physical rehabilitation and cognitive training of the minors in situations of illness. It is also a tool for health promotion, where through social games and physical education patients can learn, exercise and interact with others. It is also a tool for professionals (therapists, educators, and psychologists) can create exercises adapted to the particular needs of each patient or user groups and monitor their evolution.

The power of TANGO:H lies in its capacity to generate exercises, i.e. it is not a static platform in which exercises or games are fully defined and integrated, but also allows the implementation of these through an editor that makes this task simple. The program is capable of interpret and execute the exercises previously created by the professional in the editor TANGO:H Designer (Tangible Goals: Health Designer). The interface for the end user, is an active video game, where the patient performs the exercises previously created as a game, interacting with the system through body movements and gestures. The combination between editor and game modules allows the creation of a variety of exercises, personalized and adapted to the characteristic of the patients.

To understand the composition and elements of the exercises you need to know a number of concepts that define the interaction with the application, as described in Table 1.

In order to consider a target as reached, the user will need to interact with it with any or all of the associated contact points whether it is a target with OR logic or AND logic respectively. Additionally, a special type of target can be assigned, the Dummy that should not be reached by the user.

Table 1 Concepts and interaction elements in TANGO:H

Element	Description
Contact point	Represents a point of human body which allows user interaction with a target. The system currently has a total of 13 contact points enabled.
Objective	This is the element that the user must meet one or more contact points. A Target consists of an image, or a region of the screen, to which accompanies a set of properties: a) Contact Point. Can have one or many; b) Sound. Plays when a contact point reaches the target; and c) Colour. In TANGO:H represents the point of contact with which must be reached the Target.
	A Target has associated one or more contact points, the interaction between them responds according to one of the three following behaviours: a) All at once. All contact points should simultaneously reach the Target; b) One. At least one of the selected contact points must reach the Target; c) Dummy. This is a target that, although being reached, does not change the dynamics of the exercise, so it is not necessary define it Contact Points.
Stage	A Stage is a group of targets. To overcome a stage, you must achieve all the objectives that comprise as: a) Synchronous. The user must reach all the targets of the stage simultaneously; and b) Asynchronous. The user must reach all the targets of the stage regardless of the order or the time instant in which it occurs.
Step	A Step is a grouping of stages. To overcome a step, the user must complete the stages that compose it as: a) Sequential. The user must overcome Stages in the order in which they were created; and b) Randomized. The user must overcome Stages regardless of the order.
	On the other hand, a step can be repeated as often as deemed necessary. Also you can assign a sound, so that, at the beginning the step plays itself.
Exercise	An Exercise is a set of steps that are executed sequentially in the order in which they are defined. For the user to achieve an exercise, it must satisfy: a) All sequentially steps or all phases grouped in each of the steps (sequentially or randomly); and b) All the targets that are grouped in each of the stages (synchronously or asynchronously).
	The visualization of the exercises in screen is done by Steps. That is, the targets that make a step will be presented simultaneously on the screen.

The exercises will be displayed on screen step-by step, all targets belonging to a step will be shown on the game screen simultaneously. Once all step phases have been successfully completed, all targets will be replaced by the targets of the following step.

Using the established logic, the system classifies exercises in three different types: physical, cognitive and free. Each exercise class is considered and evaluated differently at execution time.

In the physical exercises, the professional desires the user to perform a series of specific movements, making him to reach certain targets with one or more contact points. A large number of visual hints are required in order to communicate to the user the next movement to perform as intuitively as possible. The chosen method to indicate the user the next movement was to match the target and the points of contact highlighting them with the same colour. Furthermore, this exercise class requires a sequential structure that will allow the therapist to orchestrate the exercise at editing.

Moreover, with cognitive exercises, the educator is interested in avoiding visual hints that can give away the next target to be reached. Thus, for this exercise class, a set of targets will be presented making the user to engage in a cognitive task such as relate the sound of a cat with its visual representation (matching). Through the use of sound hints, the user will know the next target without making it too obvious. Cognitive tasks do not required a pre-establish order to reach the targets.

In addition, the "free configuration" exercise type was added. This class allows the professional to create exercises on the editor ignoring any type of consideration established by the two previous types.

2.1 Game Modes

TANGO-H offers two game modes: a) single or b) multiplayer. In single mode the exercise will be conducted by one player in the categories described above (physical, cognitive or free). Besides the traditional way of playing with a single user, in the multiplayer game mode two people can play sequentially or simultaneously, both competitively or collaboratively. This last mode has been possible by the functional detection of two human bodies concurrently.

In the sequential multiplayer mode, after selecting the game, the two players performing the same exercise of equal complexity, one after the other.

However, in the simultaneous multiplayer mode, players will face the selected exercise simultaneously, working either in resolution as competing to reach as many points as possible. So, the competitive type shows the score for each player for the exercise performed, while in the collaborative multiplayer mode, users must work together to achieve the objectives, and the two users have the same score, time and stars.

3 Gamification

Regarding to the "addictive" or "engagement" component of games, we can found the "gamification" concept [7]. Gamification works to satisfy some of the most fundamental human desires: recognition and reward, status, achievement, competition & collaboration, self-expression, and altruism. Gamification taps directly into this. The game mechanics can be of different types, such as: a) behavioral (focused on human behavior and psyche), b) feedback (related with the feedback loop in the game mechanic) and c) progression (used to structure and stretches the accumulation of meaningful skills). There are other game mechanics that can be used for gamification materials and educational activities, such as: time (the players have some limited time to perform a task) or challenges between/among users (players can challenge each other and compete for the achievement of objectives, objects, medals, etc). It is also important to have other people with whom to compete, collaborate and compare accomplishments. As a general rule, humans want to interact and compete with others. When you get users to compete and collaborate as part of something bigger, it increases the stakes, adds another level of accountability and is a dynamic motivator. So, in team games must be considered separately the mechanics that influencing the team (win projects, group scores, etc.) as well as the mechanics that influencing the individual (motivation, positive reinforcement, etc.). In a best-practice implementation, a user's individual achievement should be rolled up under the group or team's success and highlighted in inter and intra group leader boards. Table 2 shows the gaming elements implemented in TANGO:H.

Table 2 Gaming elements

Elements	Description
Points	It is the most used mechanical; points are a running numerical value for a single action given or a combination of actions. In real life we handle sports scores, grades in school, etc. We reward or punish through the points given or removed, respectively.
	In TANGO-H user can get points for each exercise performed depending on the amount of the objectives achieved and the time spent to reach them. Each one of the exercises has a maximum score directly related to the solutions that the user must found during exercise. The score will be different depending on the game mode, the users are: a) In sequential type (single or multiplayer): A score for the single user; b) In simultaneous multiplayer competitive mode: Two scores, one per user; and c) In simultaneous multiplayer collaborative mode: A single score for the two users.
Leader boards	It exploits the social component; the effort is compared with other users and / or other types of classifications (global, local, etc). Leader boards give users the feeling of "fame" and "status." They also give users the chance to compete and compare with other members or players.
	This component matching and classification is exploited through different game modes and can display the leader boards in single or group. In competitive mode can also be seen the leader boards at the level of the classification exercise.
Status	Status is the ranking or level of a player, related to the scores obtained by users. Usually the users are motivated to achieve a high status.
Feedbacks	People are used to receiving feedback on their actions, it is important to reward positively and provide information to the user about his condition, the environment, and their achievements. For example, showing the progression in which the success is granularity displayed and measured through the process of completing tasks. Or giving rewards to motivate users: points, badges, trophies, virtual items, unlock able content, digital goods, etc.
	Real time scoring: The score is a real-time feature of TANGO:H that allows users to know at all times what is being evaluated for the exercise. The score is shown in the bottom centre of the screen.
	Success effect: The platform TANGO:H provides a positive feedback to players. This effect is shown when the users end a phase of the exercise that they are performing. The 'success effect' is represented as a shaded glow that fills the screen in blue and gives the user a visual positive signal.
Achievements	Achievements are a virtual or physical representation of having accomplished something, usually considered "locked" until the user have met the series of tasks are required to that "unlock" the achievement. For example virtual coins, medals or badges. In TANGO:H players get stars as rewards. A player can get the maximum amount of stars if you reach the highest score possible.

Table 2 (*continued*)

Elements	Description
Levels	The levels are related to the user experience or level of expertise (expert users, beginners, etc). They are to shorthand indicator of status in a community and show that you should be afforded respect for your accomplishments. In order to make rehabilitation exercises adaptable to the different users, we can define in TANGO:H different difficulty levels depending on the characteristics, skills and the evolution of each user. TANGO:H has several configurations parameters to customize and adapt the exercises to different users, such as: timing, action range, and target resizing.
	Timing: TANGO:H Designer includes the functionality to assign a time to the objectives of an exercise, in order to the user must have action on these objective for the stated time. This timing will be the same for all targets that make the exercise. The time parameter defined from the TANGO:H Designer is included in the XML definition of exercise. Furthermore, it can also particularize timing with a factor that is assigned to the user. The factor is defined from the user form in TANGO:H and stored in the user information database. All these settings for timing, both the exercise and in the user's profile, is evident to the user through a clock that appears when touch an object. The clock is filled continuously until completion defined time while executing the exercise. This is when the goal reached is given as valid.
	Action range: TANGO:H Designer includes the functionality to assign an action range to the objectives of an exercise. This action range shall be equal for all targets that make up the exercise, and requires the timing described in the previous section to detect whether or not the user is willing to play or is a step movement. The action range parameter defined from the TANGO:H Designer is included in the XML definition of exercise. It may also particularize this range with a factor that is assigned to the user. All these settings for action range, both in the performance and in the user's profile, are evident in TANGO:H to the user executing the exercise, because when the user touches within the range (not necessarily in order) start counting the timing clock. This functionality can also create difficulty levels for exercises and users, and enables the designer to create groups of exercises that require different skills from simple to more complex. The level can be adjusted depending on the user's progress in treatment within the possibilities and needs of each patient [5].
	Target resizing: TANGO:H Designer includes the functionality to resize an object to the size you want. This behaviour is achieved by inheriting behaviour transparent goal from the beginning was configured as resizable. This option allows the designer to create goals that are easier to reach also more difficult targets, allowing adaptation to users who have more difficulties in their movement, creating difficulty levels. This also allows professionals to work on the user's motivation, ensuring that all users will be able to achieve the objectives as appropriate. So, the designer can go adjusting the level of difficulty as the possibilities of each user.
Avatar's Customization	TANGO:H has different avatars with which the user can be identified. The goal the customization of avatar is to increase the sense of immersion and the feeling of affection for the character with which the user identifies.
User satisfaction	This feature was implemented to obtain the end-user feedback after the completion of each game on the platform TANGO:H. To determine the user' satisfaction after the execution of an exercise, at the final phase of the game a set of emoticons specially designed for the platform TANGO:H is shown. Emoticons represent different emotional categories (positive, neutral and negative). Depending on the age entered for the user, the numbers of emoticons shown vary, because the older has more emotional discrimination power.

4   Implementation and Validation

The Operating System (O.S.) employed as development platform has been Microsoft Windows 7©. When choosing the development platform, a study involving different alternatives was undertaken (drivers are available for all O.S. such as Linux). One of the main factors considered when choosing Microsoft Windows © was its compatibility with Microsoft Kinect ©. Another important factor in the decision is the large user base that this Operating System has, taking into account the objective to distribute the application to the largest number of users as possible.

The developed software has been implemented in C# on the .NET platform. The use of this platform guarantees the correct functioning of the application on Microsoft Windows © and eases the portability to future versions of this Operating System. In order to interact with Microsoft Kinect© the drivers developed by PrimeSense™ and the OpenNI™ libraries. OpenNI™ is an Open Source Framework that allows to develop applications for the Microsoft Kinect© device.

OpenNI™ allows abstracting the data obtained from the device and using them to interpret the user position and pose through the generation of a virtual skeleton. This representation consists in thirteen contact points represented by different limbs and joints of which their position and orientation can be obtained in real time.

The system development has focused in maintaining as much flexibility in exercise creation without affecting the usability of both product components. Additionally, the application has to adapt itself to the needs of the user. Therefore, it was crucial that the application was highly configurable and able to cover the cognitive needs or physical rehabilitation requirements. So, the adaptation is achieved through the logic introduced in application background responsible of executing the exercises defined in the XML.

We have followed in the design a User Centred Design (UCD) methodology [6], and validate the TANGO-H playability, usability and functionality with experts and children [3]. Note that 13 educational playability heuristics were evaluated, obtaining as educational playability average a value of 2.9, among values from 1 to 5.

Fig. 1 Validation of TANGO:H conducted with experts and children

5   Conclusions

In this paper is described the design and development of a KINECT-based interactive platform, called TANGO:H. This platform is adaptable to the characteristics of the population it is intended: hospitalized children and in home care. Also, it is highly configurable and customizable, thanks to exercises editor: TANGO:H Designer. This feature allows health professionals and educators to create game-like exercises, adapted to the specific needs of end users and to the context in which the intervention will take place. Moreover, due to the diversity of users it targets, TANGO-H is an accessible platform that allows interaction with information systems without physical contact with the traditional control systems using KINECT sensor. Therefore, this tool allows compensation for physical inactivity of children in situations of illness and aims to help young hospitalized people in their recovery and, at the same time, support health professionals and educators. Also, this tool will help to normalize children providing them quality of life and wellbeing.

Acknowledgments. This work is funded by the research project SALUD-in: *"Interactive Virtual Rehabilitation Platform based on Social Games for Health, Physical Education and Natural Interaction Techniques"* PI2010/218 of Canarian Agency for Research, Innovation and Information Society.

References

1. Alankus, G., Lazar, A., May, M., Kelleher, C.: Towards customizable games for stroke rehabilitation. In: Proceedings of the 28th International Conference on Human factors in Computing Systems, pp. 2113–2122. ACM (2010)
2. Cameirao, M., Bermudez, I., Duarte, O., Verschure, P.: The rehabilitation gaming system: a review. Studies in Health Technology and Informatics 145, 65 (2009)
3. González, C.: Student Usability in Educational Software and Games: Improving Experiences. Advances in Game-Based Learning (AGBL) Book Series. IGI Global (2012)
4. Gonzalez, C., Toledo, P., Alayon, S., Munoz, V., Meneses, D.: Using Information and Communication Technologies in Hospital Classrooms: SAVEH Project. Knowledge Management & E-Learning: An International Journal (KM&EL) 3(1) (2011)
5. Hutzler, Y., Sherril, C.: Defining adapted physical activity: internacional perspectives. Adapted Physical Activity Quarterly 24(1), 1–20 (2007)
6. Nicholson, S.: A User-Centered Theoretical Framework for Meaningful Gamification. In: Proceedings GLS 8.0 (2012)
7. Zichermann, G., Cunningham, C.: Gamification by Design: Implementing Game Mechanics in Web and Mobile Apps. O'Reilly Media, Sebastopol (2011)

Author Index